REMOTE SENSING IN LAND EVALUATION

REMOTE SENSING IN LAND EVALUATION

Dr. R. S. Yadava
Head, Department of Geography
Arunachal University, Itanagar

RAJESH PUBLICATIONS, NEW DELHI

Published in India : 1997

Published by :
M. L. Gupta
Rajesh Publications
1, Ansari Road, Darya Ganj,
New Delhi-110 002

Type Setter : **Sanjay Gupta & Km. Anita Gupta**
Sanjay Enterprises, New Delhi-110 002
Printed at : Aditya Offset Press **Delhi**

अरुणाचल राज्यपाल
GOVERNOR ARUNACHAL

राज भवन
इटानगर-७९१११
RAJ BHAVAN
ITANAGAR-791111

August 25, 1994

FOREWORD

Dr. R.S.Yadav of Department of Geography, Arunachal University has written the book, "Land Evaluation: An Application of Remote Sensing". As the title of the book suggests, it deals with systematic application of available remote sensing data of a particular land in its evaluation. The book is written in simple and lucid language.

I hope the contents of this book will be useful for evaluation of land of Lower Nizam Sagar Command area to which this study primarily relates to. It will help the planners for better utilisation of land on a long term perspective which will in turn help the farmers to make the best utilisation of their available resources; and will also improve the economic condition of the people of the area.

A number of tables and illustrations have been given in the book to facilitate understanding and evaluation of the land. I trust the book will also help students of geography, agriculture and applied disciplines. This may also encourage them to undertake similar studies for other areas of the country and useful data will be available for optimum utilisation of the land. In the final analysis, the farmers should benefit immensely from such studies.

(Mata Prasad)

CONTENTS

	Foreword	V
	Preface	IX
	Acknowledgements	XIII
	List of Figures	XV
	List of Tables	XVI
1.	Geoenvironmental Background	1
2.	Soil Survey Methodology	12
3.	Relationship Between Physiography and Soils	21
4.	Land Evaluation	44
5.	Land Irrigability Classification	78
6.	Summary and Conclusion	90
	Bibliography	93
	Appendices - Pedon	95
	Index	119

PREFACE

Agriculture is the mainstay of the people and has its direct bearing on the life of mankind. Soil is one of the natural resources which has direct and continued impact on agriculture. Therefore, it is essential to know about the soil potentiality and its suitability for various purposes, hence soil survey becomes necessary. The study of landscape helps us in identifying the soils. Soil survey and its efficiency depends largely upon a good understanding of the relationship between physiography and soil genesis. The systematic photo-interpretation technique provides excellent opportunity to study such relationship which is oriented towards physiography-cum-soil boundary.

The standard soil survey can be used for regional planning, study of soil limitations, potential areas for detailed survey, soil conservation, soil and water management, etc. A semi-detailed soil survey was conducted in a part of Nizamsagar Command Area, Nizamabad District, Andhra Pradesh, for land use planning. Soils were classified upto family level according to Soil Taxonomy (USDA, 1975). The phases of soils were also given proper emphasis, which are essential for crop land planning and management.

The survey area is characterised by high density of population and very high growth rate of population. About 70% of its population depends on agriculture and allied activities.

The present survey work was taken up for modernisation of agriculture in the command area by adopting best possible land use and agronomic practises on the basis of suitability of land. The collection of soil data aims at the interpretation of irrigable areas, soil and water management, land appraisal and development. The soil suitability for different land utilization types and irrigability have been worked out to evaluate irrigation needs of the area, according to the suitability for a particular crop. This is vital for perpetual maintenance of soil fertility, checking soil degradation, sustained soil fertility, maintenance of man-supporting capacity of land, rational utilization of these natural resources for sustained economic welfare of the people in general, and preservation of environment along with socio-economic development and human welfare.

The objective of the soil survey was to map soils as association of families and evaluate their suitability for various land utilisation types as per F.A.O. Framework of Land Evaluation (1976). It involves execution and interpretation of basic soil survey data for specific land uses. Four categories of suitability have been recognised in decending order, i.e. land suitability order, classes and sub-classes of the land units reflecting the kind of suitability within the order and kind of limitations within the classes. Rating of land qualities for four well-defined Land Utilization Types (LUTs), namely Paddy (LUT - I), Sugarcane (LUT - II), Jowar - Maize Millets, etc. (LUT - III) and Turmeric-Caster (LUT - IV) have been worked out.

This was followed by an appraisal of land qualities for all the mapping units in terms of the soil texture, slope, organic matter, soil depth, erosion and risk of flooding. It was followed by an assessment of land qualities for all the four LUTs by comparing its requirements with the characteristics of the mapping units of the area. Thus, all the mapping units were classed into highly, moderately and marginally suitable, and not suitable for various LUTs.

The highly suitable mapping units for defined LUTs were those which did not have any significant limitation to sustained application of a given use. Moderately suitable units were those which beared common limitations of soil depth, relief, risk of flooding and cannot be improved. While the limitations of fertility, PH, organic matter and erosion hazards are temporary and can be improved by adding manures, fertilizers, chemical amendments and adopting suitable soil conservation measures. The marginally suitable units have some permanent limitations such as depth of soil, relief and risk of flooding. The major limitations are fertility, organic matter, erosion hazards, which can be improved and turned moderately into highly suitable for specific land utilization types.

The not suitable units are those which cannot afford sustained application of any land utilization types due to some of the permanent limitations such as relief, soil depth, risk of flooding and steep slope. It was very much useful to indicate whether existing land use is proper or otherwise. Besides this, it helps to find out better alternative to the existing LUTs which would generate higher economic return as compared to the prevalent one.

The suitability of land for irrigation depends on geoenvironmental and socio-economic factors in addition to the soil irrigability classes. The more important considerations were the quality and quantity of water, drainage requirements and other economic considerations. Definition adopted for classifying the soils into six Land Irrigability Classes under Semi-arid and Arid regions are as follows :

Class 1 lands are those which have no limitations for sustained use under irrigation. Class 2 lands have moderate limitations for sustained use under irrigation. The limitations of these lands are moderate permeability and drainage of the soil. Class 3 lands have severe limitations for sustained use under irrigation. The topographic limitations include complex slope and undulating terrain. Soil limitation is mainly due to erosion, poor permeability and drainage. Class 4 lands are marginal for the sustained use under irrigation because of severe limitations of soil and topography. Class 5 lands are temporarily not suitable for sustained use under irrigation due to very severe limitations of topography and very low moisture whereas Class 6 lands are not suitable for sustained use under irrigation. The very severe limitations are soil, topography and excessive drainage which confine in the hilly landforms.

Thus, the land utilization according to the suitability of soil and irrigation according to land irrigability may fulfil the needs and requirements of rural population of the area by increasing yieid per unit area without disturbing the ecological balance of the command area.

ACKNOWLEDGEMENTS

I express my sincere thanks and gratitude to Prof. D.N. Chaturvedi Ex- Vice Chancellor, Kashi Vidyapith who sponsored S-1 Course of IIRS, Dehradun. I owe deep sense of gratitude to Prof. S.L. Kayastha who has been my teacher and inspired me in the field of research and associated latest tools and techniques. To Prof. A. C. Bhagawati, Vice Chancellor, Arunachal University, for encouraging such academic pursuits.

I have the deep sense of gratitude to the scientists of IIRS, Dehradun specially Prof. D.S. Kamat, Col. H.M. Dattatreya, Col. P.Mishra, late Prof. H.S. Iyer, Prof. B.M. Singh, Prof. M.L. Manchanda, Prof. L.M. Pande, Dr. Jitendra Prasad, Dr. T. Natrajan, Shri N.D. Sharma, Dr. P.S. Roy, Dr. V.C. Jha and Shri I.C. Matieda, for their continuous support right from the basics of Remote Sensing to report writing stage.

My sincere thanks are to Prof. N. C. Gautam, Prof. Venkatratnam, Dr. B.R.M. Rao, Dr. R.S. Dwivedi, Dr. Raghavaswami of N.R.S.A., Hyderabad, Prof. R.S. Chaturvedi, Dr. A.N. Singh, Dr. S.K. Shukla, of R.S.A.C., Lucknow, Prof. R.L. Singh, Prof. H.H. Singh, Prof. V.R.Singh, Dr. K.K. Dube, Dr. T.V.Singh, Dr. R.P.B. Singh and Dr. S.P. Mishra Dept. of Geography, BHU, Varanasi for their encouragement and teachings. My respected teachers Shri Anant Bahadur Singh, Principal B.I.C. Jalalpur, Shri Paras Nath Singh, Manager and Shri Prabhu Narayan Singh, Principal Shri Dhaneshwar Singh and Shri Mangla Prasad Mishra of R.I.C. Mahareon, Pureon, Jaunpur are duly acknowledged.

I am highly indebted to Shri D.R. Singh, fellow Officer Trainee from All India Soil and Landuse Survey, New Delhi for his painstaking work of soil profile classification, interpretation of boundary, assistance in land evaluation and irrigability classification of the area without which this work may have not completed.

Some of my friends namely Dr. K.P. Mishra, Dr. R.D. Dubey, Dr. D.K. Tiwari, Shri A.K.Singh (JOSEF), Shri R.K. Singh, Dr. R.P. Yadav, Dr. K. Prasad from B.H.U. and Shri V.K.B. Sawant, Shri S. Gopi, Shri P.S. Gupta,

Shri B.S. Negi, Shri Sapu Changeja, Shri Satyanrayan Officer Trainees of IIRS are acknowledged for their friendly feelings.

I have the honour to express deep sense of gratitude to my father Shri Khetaoo Yadava and mother Smt. Bipata Devi, Late Shri Ram Surat Yadava (elder brother) and Shri Jiya Shankar Yadava (younger brother) who looked after my family and rendered all possible helps to me. I express gratitude to my wife for her constant support, without which the energy and enthusiasm needed to complete this task might not have been found.

I am thankful to my faculty friends Dr. R.C. Joshi and Shri T. Riba for their valuable contributions and Shri Mrinal Nath for cartographic assistance. Shri Gopinath, A. U. has been very prompt in typing the final manuscript. I sincerely acknowledge prompt and valuable services rendered by Shri Raashid Nepal with regard to final proof reading.

I am happy to place on record the sincerity and promptness of Shri M. L. Gupta, Rajesh Publications, New Delhi, who published the book in record time.

Ram Sakal Yadava

Date : **26th January, 1997**
Place : Rono Hills, Itanagar

LIST OF FIGURES

1. India - Index Map - Survey Area 2
2. Index Map of Part of Nizam Sagar Command - Armur and Navipet Area, Nizamabad - A. P. 3
3. Temperature Variables 6
4. Ombrothermic Diagram - Nizamabad 7
5. Physiography-Cum-Soil Map of Sample Strip, Near Yemcha Village. 25
6. Schematic Cross Section of Sample Strip Near Yemcha 26
7. Physiography-Cum-Soil Map of Sample Strip Near Anksapuram 32
8. Schematic Cross Section of Sample Strip Near Anksapuram 33
9. Physiography - Cum - Soil Map of Sample Strip Daigaon to Pipri 42
10. Schematic Cross Section of Sample Strip, Daigaon to Pipri Village 43
11. Soil map-Part of Nizamabad District, A. P. Lower Nizamsagar Command, Binda-Navipet Area 61
12. Soil Map-Part of Nizamabad District, A. P. Lower Nizamsagar Command, Navipet-Anksapuram Area 62
13. Soil Map Part of Nizamabad District, A. P. Lower Nizamsagar Command, Balkonda-Bellur Area 63

LIST OF TABLES

1. Temperature and Rainfall Data for Nizamabad 8
2. Local and Botanical Names of tree prevalent in the area 9
3. Important Crops and their Botanical names 10
4. Particulars of the Aerial Photographs 13
5. Pre-field Interpretation Legend 15
6. Details of Sample Strips 17
7. Final Physiographic Legend 18
8. Legend used for Aerial Photo-Interpretation for preparation of Soil Map 38-41
9. Categories of classification for Land Evaluation 47
10. Criteria and Rating of Land Qualities for Paddy (LUT - I) 49
11. Criteria and Rating of Land Qualities for Sugarcane (LUT - II) 50
12. Criteria and Rating of Land Qualities for Jowar, Maize and Millets (LUT - III) 51
13. Criteria and Rating of Land Qualities for Turmeric and Caster (LUT-IV) 52
14. Land Qualities of various mapping units of Lower Nizam Sagar Command 54-56
15. Rating of Land Qualities of various mapping units for Paddy (LUT - I) 57-58
16. Rating of Land Qualities of various mapping units for Sugarcane Cultivation (LUT - II) 64-65
17. Rating of Land Qualities of various mapping units for Jowar, Maize and Millets (LUT - III) 69-70
18. Rating of Land Qualities of various mapping units for Turmeric and Caster (LUT - IV) 73-74
19. Land Suitability Categories and Summary of mapping units under each category for all LUTs 75
20. Relative Suitability, Potential LUTs and Irrigability of Different Mapping units 76
21. Criteria for classifying soils into Irrigability classes in Semi-Arid and Arid Regions 81-82
22. Specifications for Land Irrigability Classes 84
23. Physiography, Soils and Irrigability Classification 86-88
24. Area under different Irrigability Classes 89

CHAPTER - 1

GEOENVIRONMENTAL BACKGROUND

LOCATION AND EXTENT

The survey area is a part of Nizamsagar Command Area which occupies north-western part of Telengana pediplain. It lies between 18^0 46'N. to 18^0 55' N. latitude and 77^{0}51'E. to 78^{0}27'E. longitude having a length of 60.1 km. in east-west and 16.5 km. width in north-south. This area is covered by topographical sheets 56 F/13, 56J/1 and J/5 published by Survey of India. Survey area covers 79,668 hectares which is bounded by Haldi *nadi* in the west, the Manjra and the Godawari river in the north-west, and the Pedda *vagu* in the east. There is no physical boundary in the north and south of the area. It is expanded from Naleshwar to Ranjerla in the North and Nila to Yelpur village in the south. Administratively it occupies northern part of Nizamabad district which is surrounded by Adilabad district in the north, Karimnagar in the east, Medak in the south and the Maharashtra state in the west (Fig. 1 and 2).

GEOLOGY

Geologically, the area forms a part of ancient and stable Deccan block. Though there are variable geological formations, like dykes, residual hills, etc. but it is mainly composed of gneissic-complex of the pre-Cambrian origin with granitic intrusions in the form of dykes. The western part of the area is covered by clay alluvium, from the Godawari and the Manjra river, but the basement is primarily composed of gneiss and granitic rocks. The clay alluvium which is derived from the Deccan plateau, has been brought and deposited by the Godawari river. Some limestone tracts are present and spread in north-south direction.

PHYSIOGRAPHY AND RELIEF

Physiographically, the area has been divided into hills, pediment plain, buried pediment plain, alluvial plain and limestone landscape. The present landscape helps to apprehend various physiographic processes responsible for its present expositions. During the field work, it was

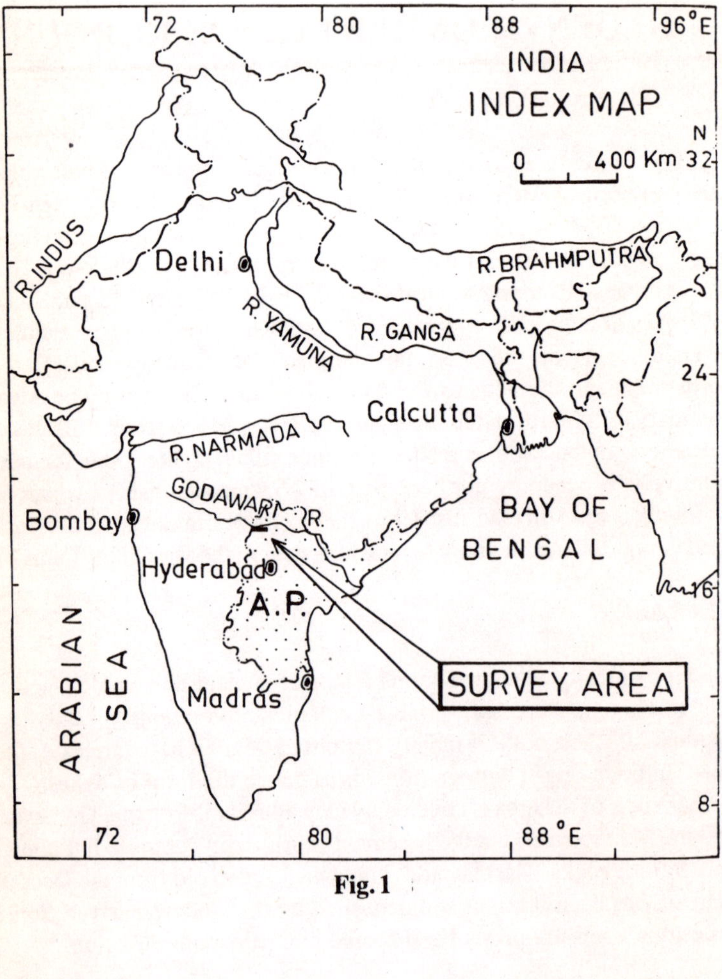
72
80
88
96°E
INDIA
INDEX MAP
N
0
400 Km
32
R. INDUS
Delhi
R. BRAHMPUTRA
R. YAMUNA
R. GANGA
24
Calcutta
R. NARMADA
GODAWARI R.
Bombay
BAY OF
BENGAL
Hyderabad
16
A.P.
ARABIAN
SEA
SURVEY AREA
Madras
8
72
80
88°E

Fig. 1

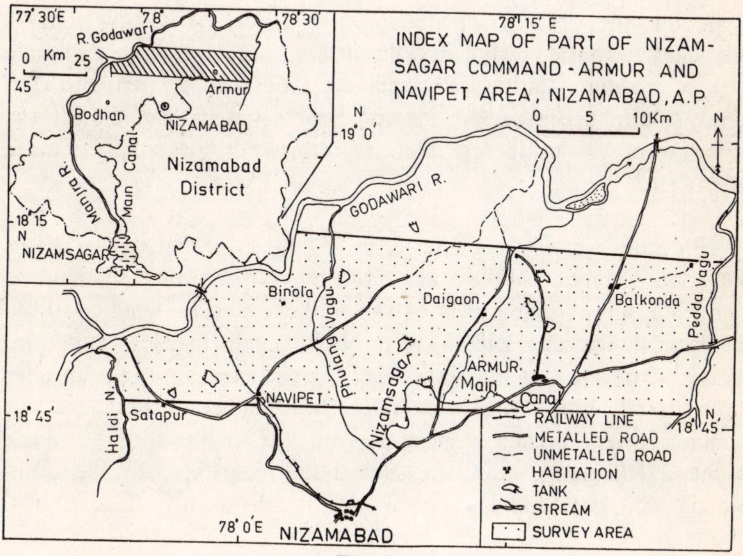
INDEX MAP OF PART OF NIZAM-SAGAR COMMAND - ARMUR AND NAVIPET AREA, NIZAMABAD, A.P.
77°30'E
78°
78°30'
78°15' E
R. Godawari
0 Km 25
45'
Armur
Bodhan
NIZAMABAD
Nizamabad District
Manjra R.
Main Canal
18°15' N
NIZAMSAGAR
N 19°0'
0 5 10Km
N
GODAWARI R.
Binola
Daigaon
Balkonda
Pedda Vagu
Phulang Vagu
Nizamsagar
ARMUR Main Canal
NAVIPET
Satapur
Haldi N.
18°45'
18°45' N
78°0'E
NIZAMABAD
RAILWAY LINE
METALLED ROAD
UNMETALLED ROAD
HABITATION
TANK
STREAM
SURVEY AREA

Fig. 2

noticed that weathering is very much active which has led wear and tear of hard rock masses like gneisses and granites. It is characterised by semi-arid climate and it receives a limited amount of rainfall during the months of summer monsoon. The rainfall washes away the weathered material from the hills and pediment plain and deposits in the depressional areas and valleys. Thus, the backward movement of slope has become an important feature of pediplained landscape. Some residual hills namely Armur and Balkonda are devoid of soil and vegetation which are merely heaps of granitic boulders. Erosion, transportation and deposition have also played significant role in the carving out of present topography. Fluvial processes are responsible for exposure of deeply buried pediment. The pediment plain can be divided into crest, side slope and valley areas. The pediment valley differs sufficiently from the hill valley and buried pediment valley in terms of colour, texture, organic carbon, moisture status, soil depth, etc. Above all, its gentle to very gently sloping pediplained landscape, punctuated by isolated hillocks in the form of inselbergs, that resembles the last phase of cycle of erosion.

Elevation varies from 330 m. to 453 m. in the area. The lowest elevation was found in the north, while highest on Yallama Gutta hill. Elevation of Armur hill is 430 m. followed by Balkonda 410 m., Yemacha 408 m. and Binola 392 m. Near Gandhi nagar, Kamalapur, it is 390 m. and 410 m. respectively. The term *Konda* stands for isolated relic hills which are merely heaps of granitic boulders. This feature is very much associated with the Andhra culture as name of many settlements close to such hills is suffixed by the word Konda such as Balkonda, Namlekonda, etc. to identify their original abode.

DRAINAGE

The area is drained by the Godawari river and its tributaries, namely, Manjra river, Pedda *vagu* and Phulang *vagu*. The Manjra river which flows towards the north, joins the river Godawari near Kankurti village. Phulang *vagu* and Pedda *vagu* also flow northward parallel to each other and join the main river near Kondur and Sarel villages respectively. These rivers are rain and spring fed. Hillocks have helped very well in construction of earthen irrigation tanks which is the special feature of the area. Hills and pediment soils are moderate to excessively drained while buried pediment plain and alluvial plains are either moderately or poorly drained.

CLIMATE

The area is characterised with the tropical monsoon type of climate but accompanied with prolonged dry periods. On the basis of local rainfall and temperature conditions, four distinct seasons may be identified. The seasons are :

(i)	South-west monsoon	(June to September)
(ii)	North-east monsoon	(October to December)
(iii)	Winter period	(January to February) and
(iv)	Warm summer	(March to May)

The summer monsoon starts with a sudden outburst of clouds in second or third week of June which brings down the mean maximum daily temperature by 3^0 to 4^0 celsius. Rain helps in providing relief from the hot winds.. The intensity of rainfall gradually increases until it becomes maximum in the month of July. The 85.6% of the annual rainfall is received during the months of south-west monsoon. The normal date of the onset of monsoon is 6th June whereas 15th October is the date of its retreat (Fig. 3).

The beginning of north-east monsoon follows a transitional period in between summer and winter season. A reversal of wind direction takes place from that of summer monsoon. Coolest month is December though temperature never falls below 16^0 C. and the average rainfall is approximately 81 mm. during this season.

The diurnal temperature starts rising from the mid-January but it never crosses 32^0 C. till the end of February. Usually sky remains calm and clear during this period while relative humidity in atmosphere continues to be very low.

The pre-monsoon period experiences a sharp rise in temperature. In the month of April it reaches approximately 39^0 C. and oscillates around it in the month of May. Scorching sun and dry warm winds prevail during this period which makes this season quite uncomfortable. This also inhibit outdoor works. However, occasional thunder showers lower the temperature considerably. The pre-monsoon season is comparably dry, though December to January are the driest months. From February to middle of May, except the occasional showers, it remains partially dry. The prolonged dry period (November to May), directly reflects the lean agricultural activities over the area. Except the months of March, April and May, when the weather becomes dust and enervating, the rest of the period is tolerable and pleasant. (Fig. 4)

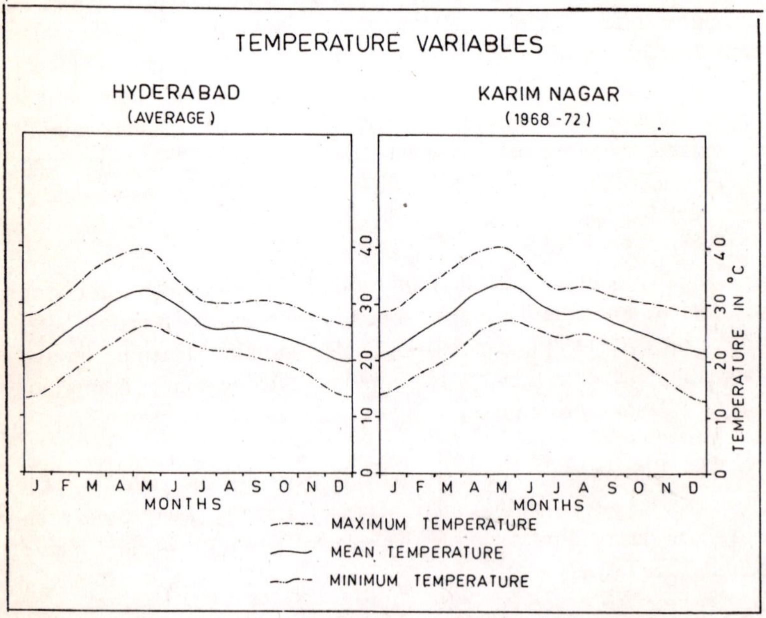
TEMPERATURE VARIABLES
HYDERABAD
(AVERAGE)
KARIM NAGAR
(1968 -72)
0
10
20
30
40
TEMPERATURE IN °C
J F M A M J J A S O N D
MONTHS
MAXIMUM TEMPERATURE
MEAN TEMPERATURE
MINIMUM TEMPERATURE

Fig. 3

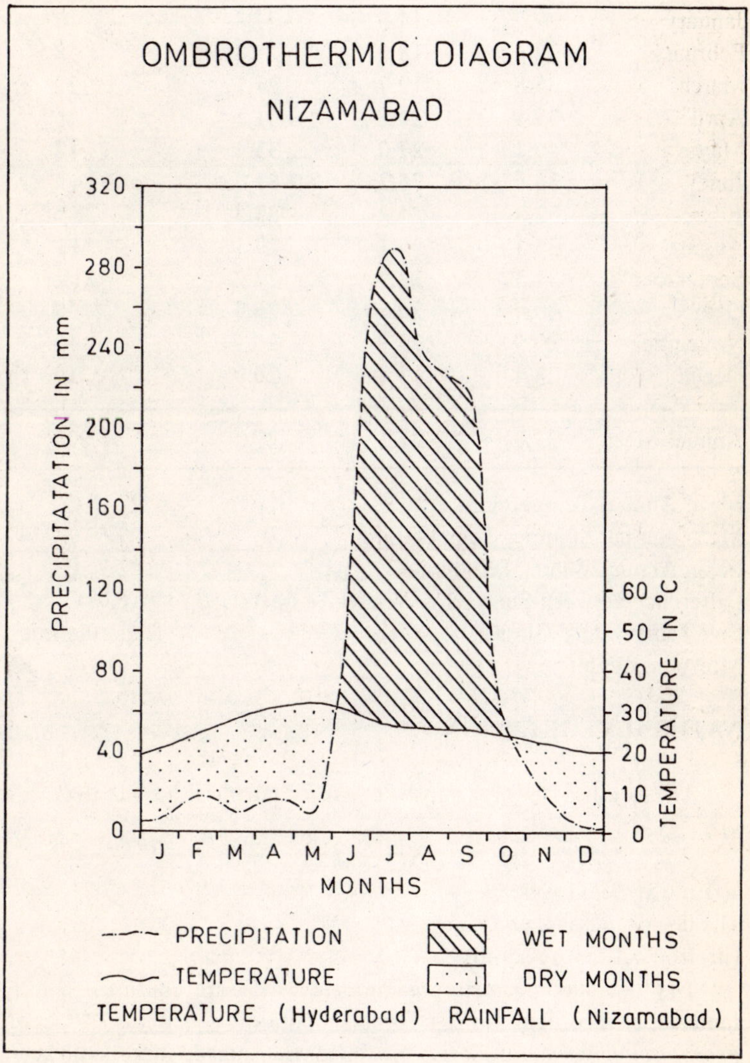
OMBROTHERMIC DIAGRAM
NIZAMABAD
PRECIPITATATION IN mm
320
280
240
200
160
120
80
40
0
TEMPERATURE IN °C
60
50
40
30
20
10
0
J F M A M J J A S O N D
MONTHS
PRECIPITATION
TEMPERATURE
WET MONTHS
DRY MONTHS
TEMPERATURE (Hyderabad) RAINFALL (Nizamabad)

Fig.4

Table 1
Temperature and Rainfall Data for Nizamabad

Month	Air temperature in ^{0}C			Rainfall
	Mean max.	Mean min.	Average	in mm.
January	29.2	14.5	21.8	6
February	32.6	18.0	25.3	18
March	35.8	19.7	27.7	12
April	39.1	24.7	31.5	16
May	40.1	27.0	33.4	12
June	36.6	26.2	31.4	147
July	32.8	24.2	28.5	288
August	33.1	25.1	29.1	233
September	31.5	22.9	27.2	219
October	30.4	20.3	25.3	54
November	29.9	16.7	23.3	17
December	28.4	13.4	20.9	10
Annual Mean	33.3	21.1	27.2	1,025

Mean Annual Temperature	:	27.2^0 C
Mean Annual Summer Temperature	:	29.6^0 C
Mean Annual Winter Temperature	:	22.7^0 C
Difference between Summer & Winter Temperature	:	6.9^0 C
Soil Temperature Class	:	Hyperthermic
Moisture Regime	:	Ustic

NATURAL VEGETATION

The forests of the area may be classified into three broad categories, viz.

(i) moist deciduous;
(ii) dry deciduous; and
(iii) tropical thorny scrubs.

Dry deciduous forests predominate which are confined to hilly landscape. Tropical thorns forests containing *acacias Albizzia-prosopissipicigera* and *cassias* are found widespread. These scrub lands are in fact transitional regions which may be termed grasslands and pastures. A secondary plant succession has taken place in the area due to deforestation, soil erosion, paucity of rainfall, overgrazing etc. Common species of vegetation of the area are :

Table 2
Local and Botanical Names of tree prevalent in the area

Local Name	Botanical Name
Dhaora	*Anogeissus Latifolia*
Mahua	*Madhuca Indica*
Tendu	*Diospyros Melanoxylon*
Amaltas	*Gassia Fistula*
Dhak	*Butea Monsperma*
Khair	*Acacia Catechu*
Imli	*Tamarindous Indica*
Mango	*Mangifera Indica*
Amla	*Emblic Officialis*
Palm	*Phaenix Sylvestris*
Neem	*Azadivachta Indica*
Sitafal	*Anno Squemos*
Rusa	*Cympobogan Martinii*
Ber	*Zizyphus Zuzuba*
Kusal	*Heteropogan Contortus*
Dub	*Cynodon Dactylon*

AGRICULTURE AND LAND USE

Agriculture is the mainstay of economy of the area. More than 70% of the population depends upon agriculture. Topography, infertility of soil, low rainfall, etc. are important constraints on cultivation. People are doing well with their perceptual knowledge regarding cultivation. Buried pediments are under intensive paddy cultivation as sufficient tank water is available. The pediment plains are sown once in the year and give poor production per hectare due to inadequate water for irrigation. Despite limited ground water, farmers manage to grow cash crops like turmeric, caster, etc. which require less water. In Armur taluka 36.1% of the area is under cultivation. Cultivable waste land account for 9.9% while uncultivable waste land account for 10% of the area. Pasture lands comprise 6.9%, current fallows, 2.8% and, new and old fallows 4.1% of the taluka. The farm size and land holding characteristics are noteworthy. About 51% of the land holdings are of less than 1 hectare, 20% are 1 to 2 ha, 16% in between 2 and 4 ha. while rests are of more than 4 ha.

Paddy is the most important crop of the area which occupies highest proportion of its gross cultivated area. Jowar, maize, pulses and chillies are also grown to considerable extent. Condiments and spices are sown in 5.00% of the total cropped area. Almost the entire agricultural activity is dependent on irrigated water. Tank irrigation predominates constituting 54% of the total irrigated area. Canals cover substantial area under irrigation. Besides, well irrigation is also significant with low energy consumption tube wells.

A considerable area is also used as grazing land as cattle rearing is an adjunct to the agricultural vocation. The cattle population is also reasonably dense as evidenced from their density.

Table 3
Important Crops and their Botanical Names

Crops	Botanical Names
Sugarcane	*Saccharum Officinarum*
Paddy	*Oriza Sativa*
Jowar	*Sorghum Vulgare*
Cotton	*Gossipium Spp.*
Groundnut	*Arachis Hypogea*
Chillies	*Capsicum annum*
Tobacco	*Nicotiyana spp.*
Moong	*Phaseolus spp.*
Bajra	*Penniselium typhoideum*

The density of livestock goes as high as 110/100 ha. of the area. There are about 70 draught animals per 100 ha. of gross cropped area. Milch cattle per 1000 population come to 99. Sheep and goat are more than 40 per 100 ha. of the cultivated area.

SOCIO-ECONOMIC CONDITION

The command area seem to be much better off than the non-command area. The farmers even spare some land for production of seeds that bring them cash returns. The land holdings are usually small and marginal, so the cultivators have to work hard for their livelihood. The working force constitute 48.07% of the total population who are engaged in varieties of vocations. Cultivators account 37.82% while agricultural labourers 18.28% , and 14.15% of the working force is engaged in house hold industries. The literacy rate in the area is also moderate. . Large settlements have favoured development of marketing facilities and recreation halls. Medical facilities are also available in towns like Armur, Balkonda, Navipet, Nandipet etc. But the small settlements area are devoid of these facilities.

INDUSTRIES AND MARKETING

The marketing facilities of agricultural products in the area are fairly adequate. The bulk of the trade is involved in products of maize, paddy, jowar, groundnut, chillies, turmeric and seeds. Most of the products are sold in the regulated markets of Armur, Balkonda and Navipet. Some of the commodities are brought and sold in Nizamabad city as it comes within a distance of 30 km. from any part of the survey area. Turmeric and tobacco processing is performed by farmers near their farm yards. There are about 20 registered factory units which employ more than 1,000 workers. Rice mills, *beedi* manufacture, saw mills, wooden furniture, tile making are also popular.

TRANSPORT AND COMMUNICATION

The study area is endowed with a good transport network. The National Highway which connects Hyderabad and Nagpur passes through the area. Armur town is located at a distance of 30 km. in the direction of north-east of Nizamabad city wherein frequent bus services are available. About 32% of the villages are linked from one another by metalled roads whereas 27% villages fall within 5 km. and rest villages more than 5 km. away from metalled road. Metalled roads are provided with public/private buses in the morning and evening. The postal services are quick in town and inefficient in the rural area. Armur is well connected with adjoining taluka headquarters and other important places in the state of Andhra Pradesh.

CHAPTER - 2

SOIL SURVEY METHODOLOGY

The aerial photo-interpretation is defined as the art and science of recognising natural and cultural features on the earth surface and concluding therefrom valuable information. The aerial photographs provide a valuable tool and the photo-interpretation is a very useful technique in mapping of natural resources data in form of thematic maps. Soils are three-dimensional natural body, having plant supporting capacity, which cannot be visualised in the photographs. Unlike forests and land use mapping, soil survey using remote sensing techniques is indirect, scientific and intelligent interpretation which requires soil studies in the field also, instead of mere photographs and mirrors stereoscope.

There are three phases of soil survey, namely, (i) soil classification; (ii) plotting boundaries of classified soil; and (iii) soil survey interpretation. The classification of soil is based on good objective description of soil profiles during the field work which is independent of the aerial photographs. The soil survey interpretation is developed from the assessment of soil properties with regard to landscape as shown on the basic soil map. Essential requirement of second phase of soil survey is a good planimetric map. The photographs provide a lot of information useful in delineation of soil boundaries accurately.

The systematic aerial photo-interpretation has been used for medium scale semi-detailed soil survey combined with sufficient field checks. Special features of this method are the use of 1 : 20,000 (approximately) photographs, detailed soil survey of sample strips and selective field checks. The concept of land systems and land facets have been used for delineation of physiographic units leading to soil map in form of physiography-cum-soil taxonomy. Soils are classified as per USDA (1975) system and mapped as association of families.

SPECIFICATION OF TOPOGRAPHIC MAPS

The topographical maps published by Survey of India were used for preparation of a base map and other uses. The number of maps used for the study are given below :

56 F	RF	1 : 250,000	Surveyed in 1966-67
56 J	RF	1 : 253,440	Surveyed in 1927-29
56 F/13 &			
56 J/1	RF	1 : 50,000	Surveyed in 1978-79
56 J/1	RF	1 : 63,360	Surveyed in 1927-28

SPECIFICATION OF AERIAL PHOTOGRAPHS

Black and White panchromatic aerial photographs on the scale 1 : 20,000 were used for both photo-interpretation and field survey. The photographs were of medium quality along with sufficient forward and lateral overlap. Specification of photographs are given below :

i) Type of Photography : Black & White panchromatic
ii) Task Number : 713 A
iii) Format Size : 23 cm. x 23 cm.
iv) Type of Print : Double weight matt
v) Scale of Photographs : 1 : 20,000 (approx.)
vi) Focal length of camera : 153 mm. (RMK)
vii) Date of Photography : 17.12.73 to 29.3.74
viii) Forward overlap : 60%
ix) Lateral overlap : 30-45%

Table 4
Particulars of the Aerial Photographs

Task No.	Run No.	Photo Nos.	Total
713 A	249	25 - 35	11
	250	1 - 28	28
	256	136 - 168	33
	259	77 - 117	41
	266	16 - 57	42
	268	8 - 43	36

Limitation : a) 60-90% side overlap in photographs of Run number 250 and 256.

b) Lateral gap between photographs of Run number 259 and 266, gap of about 100-300 m.

Mirror stereoscope (TOPCON) was used for photo-interpretation in the office as well as in field. The pocket stereoscope was used for the field work. The landscape of area was divided into hills, pediment plain, buried pediment, alluvial plain and limestone landscape. These land systems were further divided into land facets having different physiographic position in land system itself. Human influence such as land use, parcelling etc. were also taken into consideration.

SOIL OBSERVATIONS

Most of the observations were taken from soil profile pits of 50-80 cm. followed by auger boring. In order to examine the master profile, pits were dug upto a depth of 150 cm. below the surface. Soil observations were recorded on standard proforma as per USDA provided by Soil Survey Division, IIRS, Dehradun. The soil classification was done according to the 'Soil Taxonomy' (1975).

LOGISTICS

A taxi was hired for the field work and three labourers on daily wages for digging soil profile pit.

PROCEDURE

The soil survey using aerial photographs of 1:20,000 was conducted according to the methodology prescribed by VINK (1963) subsequently modified by Goosen (1967), Bennema (1969) and Shamacharya and Srinivasan (1972). Following activities were involved in completion of the project work.

PRE-FIELD INTERPRETATION

An uncontrolled mosaic of alternate photographs was prepared with the help of cellotape covered with glass plates. Aided with geologic and geomorphic information pertaining to the survey area a brief stereoscopic analysis and delineation of main land systems of the representative photographs of photo mosaic was done with the help of different photo elements. The photo-interpretation was very useful in formulation of a legend for pre-field interpretation and selection of tentative sample strips.

Preparation of photographs was done in the office itself including annotation of settlement names, principal points, conjugate principal points, minor control plots, match lines indicating effective areas of photographs bearing even numbers. The photo index was transferred on 1:50,000 topographic map for quick identification of photographic area.

Three tentative sample strips encompassing major physiographic units were selected and marked on the aerial photographs which covered all the photo analytical units supposed to have a good relationship with physiography and soils. Preference was given to areas which were situated along motorable roads. They are as given below :

i)	Hill landscape	:	Fakhrabad to Yemcha village.
ii)	Pediment plain, Buried pediment & alluvial plain	:	Anksapuram village to bridge on Pedda Vagu
iii)	Limestone landscape	:	Daigaon to Mathani village

The photographs were stereoscopically studied for analysis of topography, relief, slope and tone and a comprehensive tentative legend was framed which is given below :

Table - 5

Pre-field Interpretation Legend

H		**HILL LANDSCAPE**
	H1	Crest
	H11	Moderately dense vegetation
	H12	Sparse Vegetation
	H2	Side slope
	H21	Very steep
	H211	Moderately dense vegetation
	H212	Sparse vegetation
	H22	Steep
	H221	Moderately dense vegetation
	H222	Sparse vegetation
	H3	Foot slopes
	H31	Sparse vegetation (forested)
	H32	Cultivated
	H4	Valleys
	H41	Submerged (uncultivated)
	H42	Cultivated
	H43	Intensively cultivated
	H5	Inselbergs
P		**PEDIMENT PLAIN**
	P1	Gently undulating
	P11	Top
	P111	Forests/Scrubs
	P112	Cultivated
	P113	Waste lands
	P12	Side slopes
	P121	Terraced cultivation
	P122	Unterraced cultivation
	P123	Unterraced and uncultivated

	P13	Valleys
	P131	Submerged
	P132	Cultivated
	P133	Intensively cultivated
	P2	Flat to very gently undulating buried pediment plain
	P21	Plain
	P211	Large parcelling
	P212	Medium parcelling
	P213	Small parcelling
	P22	Valleys
	P221	Large parcelling
	P222	Medium to small parcelling
A		**ALLUVIAL PLAIN**
	A1	Gently sloping
	A11	Large parcelling
	A12	Small parcelling
	A2	Nearly level to very gently sloping
	A21	Large parcelling
	A22	Small parcelling
	A3	Severely eroded sloping lands of interfluves
L		**LIMESTONE LANDSCAPE**
	L1	Cultivated
	L2	Uncultivated

FIELD WORK

A rapid reconnaissance of the survey area was undertaken keeping in mind following objectives :

1. To find out landscape-photo-soil relationship based on pre-field interpretation.
2. To finalise the sample strips out of tentative sample strips.
3. Accessibility and trafficability of the sample area.

Fresh cuts, dugwells, etc. were examined to know geology and soils of the area. This enabled us to establish a close relationship in between landscape and soils.

SAMPLE STRIP SURVEY

A detailed study of sample strips was made one after another. Three or more observations were taken to get acquainted with the range characteristics of soils in each mapping unit like the detailed soil survey. The field investigation included the observation of minipits, auger bores and master profiles. The properties of soil profiles were recorded on the standard proforma and soils were classified upto the family level. The exact location of soil profile pits was marked on the photographs and pricked with a needle. The observation number was written towards the opposite side of the photograph and pricked points were encircled with a lead pencil. Special care was taken to mention the API unit, photo number, record physical and chemical properties of different soil horizons and classification of profile on the spot itself. Overall ninety eight observations were taken and soil samples from different horizons of master profiles were collected in sample bags for its laboratory data analysis.

Table-6
Details of Sample Strips

Sl. No.	Location	Physiographic units	No. of observations
1.	Between Fakhrabad and Yemcha village on road joining Navipet and Yemcha	H22, H31, H32 H41, H42, H5 G21, G22, G23 B12, B13, B2 A21, A22 and A3	16
2.	Near Anksapuram village on road which joins Perket with Jagtial	G12, G21, G22 G23, G32, B12 & A22	16
3.	Between Daigaon and Pipri, on road joining Armur with Nandipet	G21, G22, G23 G32, B12, B13 B2, L1 and L2	15

ADJUSTED PHOTO-INTERPRETATION

A good relationship between API unit and soil boundary was established after the detailed study of sample strips. The pre-field interpretation boundaries were adjusted according to the field observations. The tentative legend was modified as per its suitability and requirement of the area, such as the nomenclature of very gently sloping pediment plain to buried pediment plain, addition of limestone landscape after its confirmation from the lithology, deletion of unwanted or insignificant mapping units, etc. After this, alternate photographs prepared for analysis were carefully interpreted with the help of elements and techniques of photo-interpretation. Factors of site which may have influenced the tone of photographs were given proper emphasis while interpreting the aerial photographs. These factors included surface texture, relief, nature of slope, moisture condition, cultivation, vegetation covers etc. Interpretation of aerial photographs were done in three simultaneous phases :

i) Detection, recognition and identification;
ii) Analysis; and
iii) Classification of area suited for physiographic-cum-soil map.

Proper emphasis was given to terrain morphology, photo elements, ecological factors, geographic aspects and human influence in the survey area. The final legend for interpretation of photographs is given below :

Table-7
Final Physiographic Legend

H		**HILLY LANDSCAPE**
	H1	Crest with vegetation
	H11	Moderately dense vegetation
	H12	Sparse vegetation
	H2	Steep side slopes with vegetation
	H21	Moderately dense vegetation
	H22	Sparse vegetation
	H3	Foot slopes
	H31	Sparse vegetation
	H32	Cultivated
	H4	Valleys
	H41	Submerged
	H42	Cultivation
	H5	Inselbergs
	H6	Dykes
G		**GRANITIC PEDIMENT PLAIN**
	G1	Crest
	G11	Cultivated
	G12	Waste land
	G2	Side slopes
	G21	Terraced cultivation
	G22	Unterraced cultivation
	G23	Waste land
B		**BURIED PEDIMENT PLAIN**
	B1	Plain
	B11	Large parcelling (Dominantly Sugarcane cultivation)
	B12	Medium parcelling
	B13	Small parcelling
	B2	Wide valley with small parcelling

A		**ALLUVIAL PLAIN**
	A1	Gently sloping
	A11	Large parcelling
	A12	Small to medium parcelling
	A2	Nearly level to very gently sloping
	A21	Large parcelling
	A22	Small parcelling
	A3	Sloping lands of interfluves
L		**LIMESTONE LANDSCAPE**
	L1	Cultivated
	L2	Uncultivated

SELECTIVE FIELD CHECK

The photo-interpretaion-cum-soil map was complete which required post interpretation field checks to judge its validity with regard to the soil composition. Most of the observations were either from minipits or auger borings. The observation points were pricked on the photographs and numbered consecutively. Due consideration was given to areas which were located at distance or small pockets that remained unrepresented in the sample strips. Field checks were avoided from those areas which resembled identical photo pattern to the sample strips. The selective field check confirmed that there existed a satisfactory and workable correlation between the landscape and soils. Instead of taking three observations for such selective checks, attention was given to take observation from viable physiographic units characterised by similar pattern. Thus, the soil map with desired accuracy was made with higher level of confidence.

FINAL CARTOGRAPHY AND REPORT WRITING

The planimetric base map of the survey area was prepared by enlarging 1:50,000 and 1:63,360 Topographical maps on 1:20,000 scale with the help of Kargl Reflecting Projector. As there was a slight variation of scale in between the base map and aerial photographs, that instrument itself was used to transfer physiography-cum-soil boundary on the base map. Zoom transferoscope was used to transfer the rectified boundaries of sample strips on the base map. Matching of soil boundaries on different bits of base map was done while joining them together. The final cartography was done on Koda trace & the determination of area of different mapping units by dot grid method (using 1:20,000 = 1 sq.dot grid).

LAND EVALUATION

Soils are one of the most important natural resources for growing plants and crops of economic importance. The objective of the survey was to map soils and evaluate its suitability for important crops and land use as per FAO Framework of Land Evaluation (FAO, 1976). It involves execution and interpretation of basic soil survey for specific uses, i.e. paddy, wheat, turmeric, sugarcane, jowar, bajra etc. For this purpose relevant land characteristics such as soil texture, drainage, inherent fertility, electrical conductivity, relief, slope, risk of flooding were taken into account which provided sound basis for rating the land. This aims at the best use of land with limited resources while maintaining the man supporting capacity of land. Suggested land use and cropping pattern may not only curb the misuse of land but also increase per hectare yield sufficiently and favour preservation and sustenance of soil fertility in the long term perspective. This may cause socio-economic upliftment of the population and ameliorate the quality of environment.

CHAPTER - 3

PHYSIOGRAPHY AND SOIL RELATIONSHIP

The aerial photographs are very much useful for thematic mapping and assessment of natural resources such as forests, minerals, geohydrological conditions, soils, etc. Mosaics prepared from aerial photographs provide synoptic view of the area which cannot be seen at a time in the field. Three dimensional view available from photographs, when viewed under mirror stereoscope, gives an additional advantage in soil mapping as it becomes very easy to delineate physiographic/land units which are closely associated with soils. The relationship exists principally because of interaction between physiographic processes and the factors of soil formation, hence the knowledge of the physiographic processes serves to know the general pattern of the development of soil in an area.

The soil is a three dimensional natural body which cannot be characterised by surfacial features alone but surfacial features provide sufficient clues for delination of soil boundaries. Such expressions, are changes in vegetation, topography, soil colour, etc. , which are closely related to physiographic processes and changes in the soils.

Systematic approach of photo-interpretation added with sufficient co-lateral information regarding geology, physiography, climate, soils, etc. has been used to classify the area into land systems and land facets. The survey area has been divided into 5 main land types as follows :

i) Hill landscape ;
ii) Granitic pediment plain ;
iii) Buried pediment ;
iv) Alluvial plain ; and
v) Limestone landscape.

HILL LANDSCAPE

Geologically hill landscape consists of granite and granitic gneisses of Precambrian origin which have been subjected to prolonged denudation due to weathering and erosion. The present landscape resembles pediplained land forms with residual hills, very resistant to weathering and erosion. The soils of these units are well drained to excessively drained with high infiltration and run-off ratio. Ochric epipedon and cambic or argillic horizons are noticed, depending upon the catenary sequence of topography. The foot-slopes are subject to sufficient weathering and high rate of infiltration which favour for development of soil profile.

GRANITIC PEDIMENT PLAIN

It is composed mainly of granites and granitic gneisses. These units exhibit gently undulating slope, well drained, coarse textured soils and high infiltration-runoff ratio. Surface layer is characterized with light colour soil. Crest parts of the pediment plain have very thin epipedon while side-slope and valley comprise of thick ones. There was no lime reaction noticed throughout the profile.

BURIED PEDIMENT

The buried pediment is nearly level to very gently sloping area which is composed of fine alluvium brought from pediment plain. The soils are moderately well drained to imperfectly drained with sufficient moisture conditions favourable for rapid weathering and micro-biological activities. Evidences of gleyed cambic horizons, carbonate removal and mild degree of illuviation of clay have been noticed in these units.

ALLUVIAL PLAIN

The alluvial plain is formed due to fluvial action and deposition of sediments in the flood plain. These are comprised of low levees, backswamp area, eroded sloping lands of interfluves with nearly level to very gently sloping landscape of the Godawari and Manjra rivers. This landscape is subject to inundation of three to four times in the period of ten years. Soils are moderately well drained with sufficient moisture available in its subsurface horizon. Cracks and slicken sides have been noticed in the area where soils are very deep. Sub-angular to angular blocky structure has been marked in this landscape. The presence of moisture is favourable for micro-biological activities and development of soil profile. These soils are used for sugarcane and jowar cultivation along with paddy.

LIMESTONE LANDSCAPE

Narrow strips of limestone landscape elongated in north-south direction are lying almost parallel to one another. Uncultivated areas are subject to deep and wide cracks with wedge shaped slicken sides. Lime concretions in form of *Kankars* have been found. Gilgai relief has been noticed in uncultivated areas whereas cultivated ones with sufficient irrigation do not exhibit micro-relief differences. The soils are developed *in situ* and are similar to black cotton soil. These are comprised of sub-angular to angular blocky structures and montmorillonitic mineralogy. These are generally devoid of vegetation cover, occasionally small grasses cover the surface wherever soil moisture is favourable. Most often these are left abandoned and require irrigation and proper management practises for cultivation.

(H) HILL LANDSCAPE

This landscape has been divided into six sub-units based on rock composition, catenary sequence, slope, land use, etc. It occupies 5,350 ha. of land which consists of 6.71% of the surveyed area.

(H1) CREST WITH FORESTS

This is composed of granites and granitic gneisses of pre-Cambrian period. It has occupied areas of higher elevation which exhibit gentle to moderate slopes. Surface texture of the soil varies from sandy clay loam to clay loam whereas its depth ranges in between 50 to 100 cm. It has further been sub-divided into two groups based on the quality of vegetation.

(H11) Crest, Moderately Dense Forests

This landscape exhibits gentle slope (three to five per cent) and has undergone different soil forming processes such as weathering, clay illuviation, leaching of $CaCO_3$ etc. The surface texture of the soil varies from sandy clay loam to clay loam and the soil depth from 50 to 100 cm. permitting root penetration downward. Further, it is characterised of stoniness and rockiness with moderate soil erosion.

The soils of this mapping unit are classified as Fine loamy Udic Ustochrepts.

(H12) Crest With Sparse Vegetation

This mapping unit has been subjected to degraded forests due to overgrazing and clear felling practises. Soils are well to excessively drained which have experienced clay illuviation in B horizon in form of argillic horizon with high base saturation. It is characterized with gentle to moderate slope (three to ten per cent), and moderate to severe erosion. The surface texture of the soil varies from sandy loam to sandy clay loam. Extreme phases of stoniness and rockiness were prevalent in this landscape. The presence of moisture in the soil during the pre-monsoon period goes down and becomes inadequate to support plants and vegetation cover. Thus, the quality of forests occuring in this mapping unit is poor.

Loamy skeletal Udic Haplustalfs have been encountered in this landscape in association with Loamy skeletal Udic Paleustalfs.

(H2) STEEP SIDE SLOPE

The hill side-slopes are characterized with F and G slope classes (fifteen to thirty-three per cent). The physical weathering along with pedogenetic processes namely clay illuviation, leaching of calcium carbonate, etc. have taken place in the area. Soils which have developed upon the rocky base are devoid of sub-surface horizon. The lower portion of the side-slopes exhibits considerably deep soil profile with marked clay illuviation.

(H21) Side-Slope With Moderately Dense Forests

This mapping unit is subject to a moderate erosion and formation of soil. The surface texture of the soil varies from sandy clay loam to sandy loam with a soil depth of 25-50 cm. These areas exhibit R4 phase of rockiness. Soils are excessively drained along with clay illuviation in B horizon.

Fine loamy Typic Haplustalfs were encountered in association with Fine Loamy Udic Haplustalfs in this unit.

(H22) Side-Slope With Sparse Vegetation

The poor vegetation cover and high degree of slope are mainly responsible for moderate to severe soil erosion. The surface texture of the soil varies from gravelly sandy loam with moderately deep soil profile (25-50 cm). Rock exposures account for about fifty to ninety per cent of the areal expanse of the unit. (Fig. 5 & 6).

This physiographic unit has Loamy Skeletal Lithic Ustorthents in association with Fine Loamy Udic Haplustalfs.

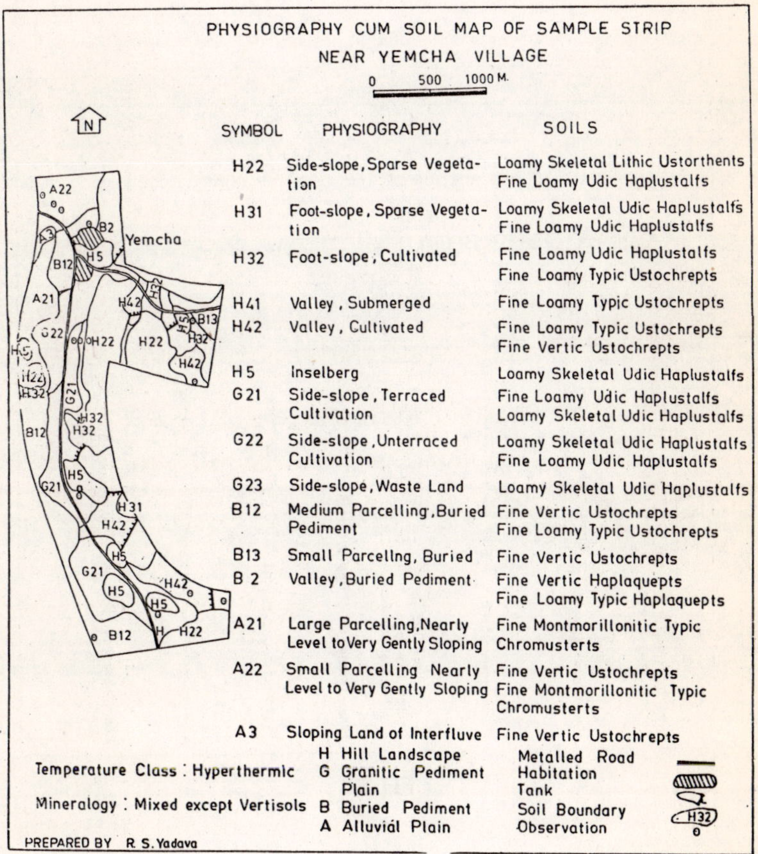
PHYSIOGRAPHY CUM SOIL MAP OF SAMPLE STRIP
NEAR YEMCHA VILLAGE
0 500 1000 M.
N
Yemcha
SYMBOL PHYSIOGRAPHY SOILS
H22 Side-slope, Sparse Vegetation Loamy Skeletal Lithic Ustorthents Fine Loamy Udic Haplustalfs
H31 Foot-slope, Sparse Vegetation Loamy Skeletal Udic Haplustalfs Fine Loamy Udic Haplustalfs
H32 Foot-slope, Cultivated Fine Loamy Udic Haplustalfs Fine Loamy Typic Ustochrepts
H41 Valley, Submerged Fine Loamy Typic Ustochrepts
H42 Valley, Cultivated Fine Loamy Typic Ustochrepts Fine Vertic Ustochrepts
H5 Inselberg Loamy Skeletal Udic Haplustalfs
G21 Side-slope, Terraced Cultivation Fine Loamy Udic Haplustalfs Loamy Skeletal Udic Haplustalfs
G22 Side-slope, Unterraced Cultivation Loamy Skeletal Udic Haplustalfs Fine Loamy Udic Haplustalfs
G23 Side-slope, Waste Land Loamy Skeletal Udic Haplustalfs
B12 Medium Parcelling, Buried Pediment Fine Vertic Ustochrepts Fine Loamy Typic Ustochrepts
B13 Small Parcelling, Buried Fine Vertic Ustochrepts
B 2 Valley, Buried Pediment Fine Vertic Haplaquepts Fine Loamy Typic Haplaquepts
A21 Large Parcelling, Nearly Level to Very Gently Sloping Fine Montmorillonitic Typic Chromusterts
A22 Small Parcelling Nearly Level to Very Gently Sloping Fine Vertic Ustochrepts Fine Montmorillonitic Typic Chromusterts
A3 Sloping Land of Interfluve Fine Vertic Ustochrepts
Temperature Class : Hyperthermic
Mineralogy : Mixed except Vertisols
H Hill Landscape
G Granitic Pediment Plain
B Buried Pediment
A Alluvial Plain
Metalled Road
Habitation
Tank
Soil Boundary
Observation
PREPARED BY R S. Yadava
FIG. 5

Fig. 5

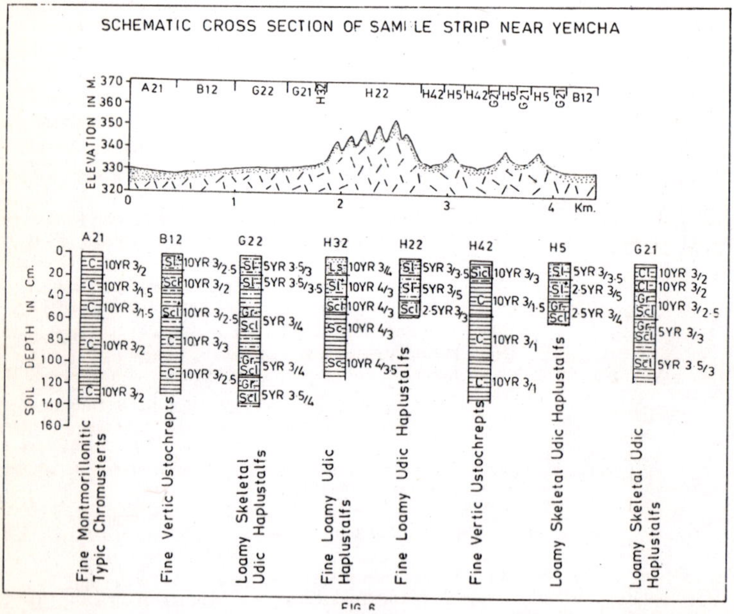
SCHEMATIC CROSS SECTION OF SAMPLE STRIP NEAR YEMCHA
ELEVATION IN M.
SOIL DEPTH IN Cm.
Km.
A21
B12
G22
H32
H22
H42
H5
G21
Fine Montmorillonitic Typic Chromusterts
Fine Vertic Ustochrepts
Loamy Skeletal Udic Haplustalfs
Fine Loamy Udic Haplustalfs
Fine Loamy Udic Haplustalfs
Fine Vertic Ustochrepts
Loamy Skeletal Udic Haplustalfs
Loamy Skeletal Udic Haplustalfs

FIG. 6

Fig. 6

(H3) FOOT SLOPES

These are the gently sloping units (three to five per cent) of the region. These are comprised of good run-off and infiltration ratio which is quite suitable for weathering and soil formation. Alteration has taken place to qualify the sub-surface as argillic and cambic horizons. This has been sub-divided as under :

(H31) Foot-Slope With Sparse Vegetation

This is closely associated with slopes of the hill landscape. Surface texture of the soil varies from gravelly sandy loam to sandy loam. Based on the depth, soils have been classed as deep (50-100 cm.). Clay illuviation has taken place in B horizon. Moderate to severe soil erosion has taken place in the area.

Dominant soil encountered in this unit was classified as Loamy Skeletal Udic Haplustalfs with association of Fine Loamy Udic Haplustalfs.

(H32) Foot-Slope Cultivated

This mapping unit is characterized with loamy sand to sandy loam surface texture of the soil. These are well drained soils which are suitable for crops like jowar, maize, etc. Soil depth varies from 50-100 cm. The coverage of solid rock mass ranges between two to ten per cent area of the unit. Slight to moderate soil erosion has taken place. Single rain-fed crops are grown during the months of summer monsoon.

Soils are classified as Fine Loamy Udic Haplustalfs in association of Fine Loamy Udic Ustochrepts.

(H4) HILL VALLEYS

The hill valleys are composed of recent and sub-recent alluvium and bounded by hill side slopes. Gneissic complex basement has been noticed immediately below the alluvium. Dominant slope of this mapping unit is one to three per cent. Only limitation of the unit is inadequate amount of water for irrigation. Bunding and terracing are the most important man-made features of this landscape.

(H41) Hill-Valley Submerged

This unit represents the submerged part of the hill valley. Surface texture of the soil varies from sandy clay loam to clay loam. Soils are very deep with a thickness of more than 100 cm. Rock outcrops occupy two to ten per cent area of this unit.

Soils were classified as Fine Loamy Typic Ustochrepts.

(H42) Hill-Valley Cultivated

The soils of this physiographic unit are characterized by sandy clay loam to silty clay loam texture in its surface layers. The soils are very deep (more than 100 cm.) and exhibit two to ten per cent rock exposure and none to slight erosion.

The dominant soil encountered in this area was Fine Loamy Typic Ustochrepts. Fine Vertic Ustochrepts were also noticed in association of the earlier ones.

(H5) INSELBERGS

The inselbergs have been formed as a result of the process of pediplantation. These are characterized with concave relief while slope ranges in between ten to thirty-three per cent. Surface texture of the soil varies from sandy laom to sandy clay loam while its depth ranges between 25 and 50 cm. Rock exposures account 25-50% of the exposed surface. Soils are very old which has permitted significant clay illuviation in the B horizon.

The soils classified in this mapping unit are Loamy Skeletal Udic Haplustalfs.

(H6) DYKES

A number of dykes have been noticed in the area which cut across the lithology of landscape. The most prominent dyke of the area starts about 500 m. away from the Armur and passes towards the north. Soils are heavy textured, varying from silty clay to clay. Depth of the soil goes beyond 100 cm. This landscape can well be assigned the phase of stony land (fifteen to ninety per cent) and rocky land (fifty to ninety per cent rock explosure). Total loss of the epipedon has been noticed in this mapping unit which is result of severe erosion. Very deep argillic horizon was exposed on the surface.

Soils are classified as Fragmental Udic Haplustalfs.

(G) GRANITIC PEDIMENT PLAIN

This landscape has been characterized by the light colour, low organic matter, well drained conditions, high base saturation, moderate infiltration and percolation, eluviation and illuviation processes, and *in situ* development of the soil. This landscape is spread over an area of 44,355 ha. which comprised 55.69% of the survey area. These soils are developed on erosional landforms namely pediments which are subject to the intense weathering, alteration and leaching of $CaCO_3$.

(G 1) CREST OF PEDIMENT PLAIN

This landscape is dominantly the result of pediplantation processes and is in the penultimate stage of pediplantation. The degree of slope envisaged in this mapping unit remain very gentle (one to three per cent). This has been sub-divided into two units as follows :

(G 11) Crest Cultivated

This mapping unit is characterized with loamy sand and gravelly sandy loam surface texture, 50-100 cm. deep, less than two per cent rock exposure. Only those crops, which require less irrigation and well drained soils, are grown in this area with the help of tube-well irrigation. Turmeric, caster, jowar, etc, are the main crops grown in this mapping unit. Slight and moderate soil erosion has taken place in such area.

The major soils in this unit are Loamy Skeletal Udic Haplustalfs and Fine loamy Udic Haplustalfs.

(G 12) Crest Waste Land

This mapping unit is comprised of loamy sand and sandy loam texture of the surface, deep soil profile (50-100 cm.), two to ten per cent rock exposure, excessively drained condition etc. These are idle lands where some scrubs, palm tree, etc. are found which are used for grazing purposes. These are subject to moderate and severe soil erosion.

Loamy skeletal Udic Haplustalfs are encountered in association with Fine loamy Udic Haplustalfs.

(G2) SIDE-SLOPES OF PEDIMENT PLAIN

The general slope of this mapping unit varies from three to five per cent. Soils are *in situ* developed and characterized with sufficient weathering, carbonate removal and well drained in nature. Clay illuviation has taken place to the extent which is sufficient to qualify for Alfisols. This landscape has been divided into three types as given below:

(G21) Side Slope, Terraced Cultivated

This mapping unit occupies larger part of the survey area. Land levelling and terracing have been observed as special feature of this unit. Terracing has been done for cultivation of summer monsoon paddy crop. The comparative heavier texture on the surface enable water stagnation, essential for paddy cultivation. Surface texture of the soil varies from sandy laom, sandy clay laom to clay loam which is moderately well drained in nature. Soil depth varies from 50-100 cm. The rock outcrops occupy less than two per cent of the surface. Paddy and maize are dominant crops grown in this area.

Soils encountered in this unit are Fine loamy Udic Haplustalfs in association with Loamy Skeletal Udic Haplustalfs.

(G22) Side-Slope , Unterraced Cultivation

This mapping unit consists of medium sized parcelling of the fields. Those are well drained soils and are used for growing crops which require light irrigation. Soil depth varies from 50-100 cm. which is subject to slight erosion. Clay illuviation has taken place to qualify the sub-surface as argillic horizon.

Soils are classified as Loamy Skeletal Udic Haplustalfs and Fine Loamy Udic Haplustalfs.

(G23) Side-Slope Waste Land

There are many limitations to use this soil for cultivation purposes. Surface texture of the soil is characterized by gravelly sandy laom and sandy loam whereas the soil depth varied from 25 - 50 cm. Rock outcrops comprise of 2-10 per cent of the exposed surface. These soils are subject to moderate and severe erosion thus it is not suitable for cultivation.

Soils are classified as Loamy Skeletal Udic Haplustalfs.

(G3) PEDIMENT VALLEY

The pediment valleys are surrounded by pediment plains. The fine materials of the soil which are washed away from pediment plain are subject to deposition in thin layers in these valleys. This landscape has further been subdivided into two types as discussed below :

(G31) Valley Partially Cultivated

This type of the valley section, which remains seasonally submerged, is cultivated occassionally. Surface texture of the area is clay laom and silty clay loam while depth of soil is more than 10 cm. Soils are moderately well drained.

Soils encountered in this mapping unit are Fine Aquic Ustochrepts.

(G32) Valley Cultivated

These are narrow flat floor valleys and characterized with fertile soil. This mapping unit exhibits clay loam to sandy clay loam texture, very deep soil profile (more than 100 cm.) and moderately well drained soil.

Dominent soils encountered in this mapping unit are Fine laomy Aeric Haplaquepts in association with Fine loamy Fluventic Ustochrepts.

(B) BURIED PEDIMENT PLAIN

This landscape is characterized with nearly level to very gently sloping plain which is covered by fine alluvium unlike pediment plain. The depth of alluvium is very deep, thus it is termed as buried pediment plain. The fine texture of horizon is suitable for stagnation of water which is necessary for paddy cultivation. Soils are moderately well drained to imperfectly drained which are responsible for gleyed cambic sub-surface horizon. Carbonate removal is also significant.

(B1) BURIED PEDIMENT PLAIN

The plain landscape is very gently sloping (one to three per cent). It is characterized by sufficient heavy textured soil while drainage is poor. This landscape has been divided into three mapping units based on the size of parcelling which bear significant relationship with soils.

(B11) Large Parcelling Buried Pediment Plain

This mapping unit is very gently sloping (one to three per cent). Soil texture of the surface varies from clay loam to silty clay loam. These soils

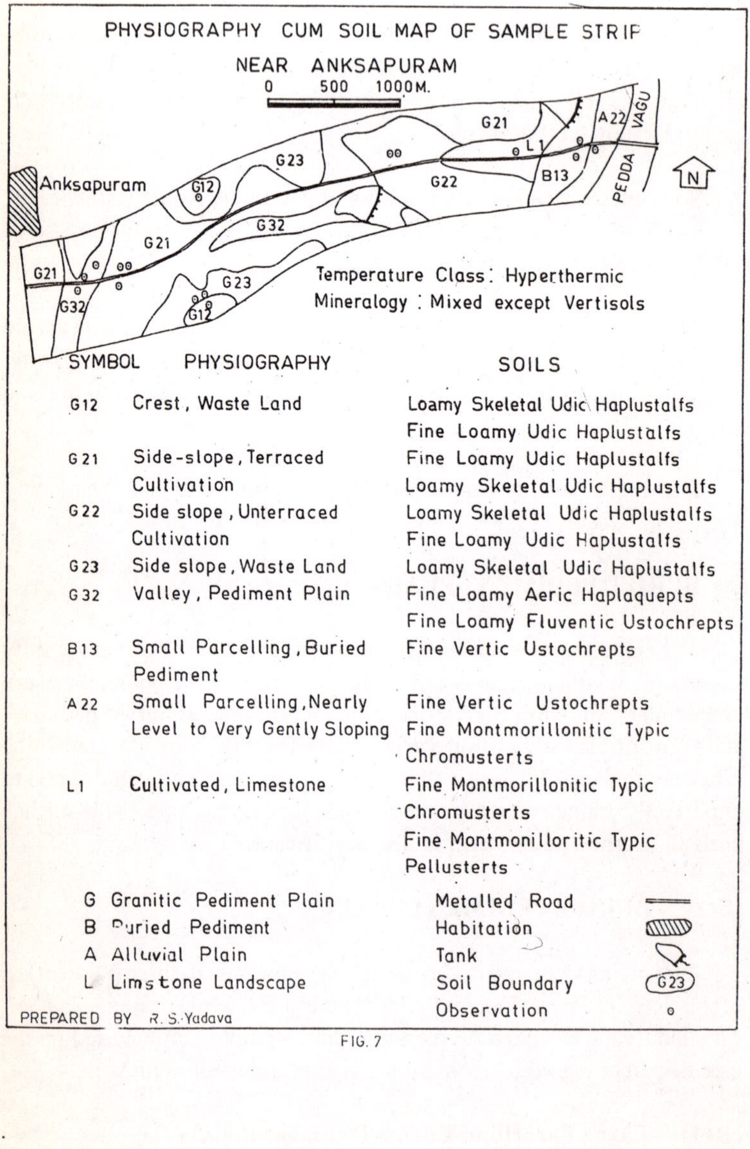
PHYSIOGRAPHY CUM SOIL MAP OF SAMPLE STRIP
NEAR ANKSAPURAM
0 500 1000M.
Anksapuram
G21
G23
G22
G12
G32
L1
A22
B13
PEDDA VAGU
N
Temperature Class: Hyperthermic
Mineralogy : Mixed except Vertisols
SYMBOL PHYSIOGRAPHY SOILS
G12 Crest, Waste Land Loamy Skeletal Udic Haplustalfs
Fine Loamy Udic Haplustalfs
G21 Side-slope, Terraced Cultivation Fine Loamy Udic Haplustalfs
Loamy Skeletal Udic Haplustalfs
G22 Side slope, Unterraced Cultivation Loamy Skeletal Udic Haplustalfs
Fine Loamy Udic Haplustalfs
G23 Side slope, Waste Land Loamy Skeletal Udic Haplustalfs
G32 Valley, Pediment Plain Fine Loamy Aeric Haplaquepts
Fine Loamy Fluventic Ustochrepts
B13 Small Parcelling, Buried Pediment Fine Vertic Ustochrepts
A22 Small Parcelling, Nearly Level to Very Gently Sloping Fine Vertic Ustochrepts
Fine Montmorillonitic Typic Chromusterts
L1 Cultivated, Limestone Fine Montmorillonitic Typic Chromusterts
Fine Montmonilloritic Typic Pellusterts
G Granitic Pediment Plain
B Buried Pediment
A Alluvial Plain
L Limstone Landscape
Metalled Road
Habitation
Tank
Soil Boundary
Observation
PREPARED BY R. S. Yadava

FIG. 7

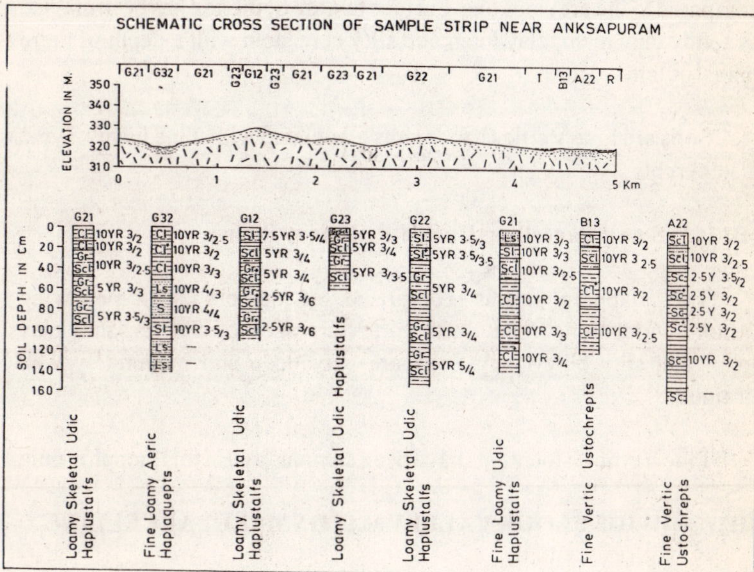
SCHEMATIC CROSS SECTION OF SAMPLE STRIP NEAR ANKSAPURAM
ELEVATION IN M.
SOIL DEPTH IN Cm
Loamy Skeletal Udic Haplustalfs
Fine Loamy Aeric Haplaquepts
Loamy Skeletal Udic Haplustalfs
Loamy Skeletal Udic Haplustalfs
Loamy Skeletal Udic Haplustalfs
Fine Loamy Udic Haplustalfs
Fine Vertic Ustochrepts
Fine Vertic Ustochrepts

FIG. 8

are very deep, moderately well drained in nature. The none to slight erosion has been witnesses in this area. These are subject to crack during dry season. (Fig. 7 & 8)

These soils are classified as Fine Montmorillonitic Typic Chromusterts with association of Fine Vertic Ustochrepts.

(B12) Medium Parcelling Buried Pediment Plain

This unit of buried pediment plain comprises of medium parcelling pattern. It is endowed with slightly lighter surface texture of soil as compared to the previous one. Surface texture of the soil has been classed as sandy clay loam, clay loam and silty clay laom with a depth of more than 100 cm.

Soils are Fine Vertic Ustochrepts associated with Fine loamy Typic Ustochrepts.

(B13) Small Parcelling Buried Pediment Plain

This mapping unit is comprised of smooth landscape having minor undulations. The surface texture of the soil is sandy clay loam and silty clay loam. Soils are very deep and moderately well drained.

Fine Vertic Ustochrepts have been encountered in this mapping unit.

(B2) WIDE FLOOR VALLEY WITH SMALL PARCELLING

This mapping unit is characterized with nearly level terrain with very deep soil. These are most fertile soils of the area and remain under permanent cultivation. Surface texture of the soil is fine. The fine clay, which is washed away from pediment plain, gets deposited in such wide valleys.

Soils encountered in this area are Fine Vertic Haplaquepts and Fine loamy Typic Haplaquepts.

(A) ALLUVIAL PLAIN

The alluvium of the Godawari river and its tributaries has been derived from basalt region through which these rivers flow. This alluvium which is clayey in texture has to pass through the alternate dry and wet seasons in semi-arid type of climate. Thus, soils are subject to alternate expansion and contraction leading to deep cracks. Padda *Vagu* and the Phulang *Vagu* alluvium is characterized by lighter surface texture as compared to Godawari one. This landscape has been divided into two parts, namely, gently sloping alluvial plain and nearly level to very gently sloping alluvial plain.

(A1) GENTLY SLOPING ALLUVIAL PLAIN

This unit is characterized with a slope of three to five per cent. Surface texture of the soil is clay in nature. Soils are very deep and fertile. This land-scape has further been sub-divided into two mapping units depending upon the size of parcelling :

(A11) Large Parcelling Alluvial Plain

This mapping unit is very significant in the upper portion of Nizamsagar Command Area and it is associated with the Manjra river alluvium. Surface texture of the soil is clayey while soil profile is very deep. This is put mainly under sugarcane cultivation.

Soils in this physiographic unit are classified as Fine Montmorillonitic Typic Chromusterts.

(A12) Small to Medium Parcelling Alluvial Plain

This mapping unit exhibits small parcelling which are suitable for wet cultivation. Surface texture of the soil is clayey in nature and moderately well drained to imperfectly drained sub-surface. Soils are very deep and subject to none to slight erosion. Stones and rock exposures are not significant to pose any hindrance for cultivation.

Soils encountered in this unit are Fine Montmorillonitic Typic Chromusterts.

(A2) NEARLY LEVEL TO VERY GENTLY SLOPING ALLUVIAL PLAIN

This landscape is situated along the bank of the Godawari river and its major tributary, Manjra river. It is characterized with heavy textured soil, thus quite suitable for paddy and sugarcane cultivation. Water is available for irrigation purposes. This landscape too has been sub-divided into two parts depending upon the size of parcelling :

(A21) Large Parcelling Nearly Level Alluvial Plain

This mapping unit happens to occur in areas closer to the Godawari river, and is quite suitable for sugarcane cultivation. This is characterized with very deep soil, moderately well drained, slight erosion, insignificant stoniness and rockiness. Due to sufficient moisture retention capacity, this soil is used for growing two crops. During the period of summer monsoon, these are used for paddy cultivation while in dry season *jowar* and *bajra* seed-crops are grown.

The soils of this unit are classified as Fine Montmorillonitic Typic Chromusterts.

(A22) Small Parcelling Nearly Level Alluvial Plain

This mapping unit has been found in the Pedda *Vagu* and Phulang *Vagu* alluvium. These are characterized by very gently sloping landscape with slightly convex slope. Surface texture of the soil varies from sandy clay loam to silty clay loam. The depth of soil goes to more than 100 cm. wheras traces cf slight erosion have been marked occasionally. Rock outcrops confine to less than two per cent of the exposed surface. Sufficient amount of water is available for irrigation. The ground water, being very close to surface, is tapped by shallow tube-wells.

Soils encountered in this unit are classified as Fine Vertic Ustochrepts.

(A3) SLOPING LAND OF INTERFLUVE

This mapping unit consists of severely eroded land. The slope varies from three to five per cent, while surface is covered with gravels and lime concretions along with rills and gullies. This unit requires soil conservation practices to check the gullies advancing any more. Surface cover, gully plugging and bunds against gullies to arrest silt and clay material need to be practised in such units.

Soils encountered in this unit are classified as Fine Vertic Ustochrepts.

(L) <u>LIME STONE LANDSCAPE</u>

This landscape is comprised of limestone lithology, having a clear unconformable relationship with its underlying granitic terrain. Soils are characterized with *in situ* formation in this area. Cracks and gilgai relief are the most important features of this area. This unit has been subdivided into two mapping units which are mainly based upon land use.

(L1) CULTIVATED LIMESTONE LANDSCAPE

The surface texture of soil in this mapping unit is clayey. It is characterized with moderately well drained and very deep soils. Currently this landscape is under paddy cultivation. Slight erosion has been noticed in this area. Soils are subject to cracks during the dry winter season and put hindrance for irrigation and cultivation. Slope ranges from one to three per cent. Wedge shaped intersecting slicken sides are well developed upto a depth of 100 cm. (Fig. 9 and 10)

Fine Montmorillonitic Chromusterts have been encountered in association with Fine Montmorillonitic Typic Pellusterts in this landscape.

(L2) UNCULTIVATED LIMESTONE LANDSCAPE

Due to semi-arid climate, cracks developed in this landscape are deep and wide enough to make irrigation impracticable. Sink holes and gilgai micro-relief and slicken sides on wedge shaped natural structural aggregates have been noticed as special features of this mapping unit. These are subject to moderate erosion. Soil depth goes more than 100 cm. Inadequate water supply is major problem which restricts its use for cultivation of crops that require irrigation. (Table 8)

Soils encountered in this landscape are classified as Fine Montmorillonitic Type Chromusterts.

Table 8
Legend Used For Aerial Photo-Interpretation For Preparation Of Soil Map

Symbol	Physiography and Land Use	Soil Association	Surface Texture
1	2	3	4
H	Hills		
H1	Crest with Forests		
H11	Moderately Dense Forests	Fine Loamy Udic Ustochrepts	Scl, Cl
H12	Sparse Vegetation	Loamy Skeletal Udic Haplustalfs, Loamy Skeletal Udic Paleustalfs	Sl, Scl
H2	Side Slope With Forests		
H21	Moderately Dense Forests	Fine Loamy Typic Haplustalfs	Sl, Scl
		Fine Loamy Udic Haplustalfs	
H22	Sparse Vegetation	Loamy Skeletal Lithic Ustochrepts	Gr, Sl
		Fine Loamy Udic Haplustalfs	Sl
H3	Foot Slope		
H31	Sparse Vegetation	Loamy Skeletal Udic Haplustalfs	Gr, Sl
		Fine Loamy Udic Haplustalfs	Sl
H32	Cultivated	Fine Loamy Udic Haplustalfs	Ls, Sl
		Fine Loamy Udic Ustochrepts	
H4	Valley		
H41	Submerged	Fine Loamy Typic Ustochrepts	Scl, Cl
H42	Cultivated	Fine Loamy Typic Ustochrepts	Scl, SiCl
H5	Inselberg	Loamy Skeletal Udic Haplustalfs	Sl, Scl
H6	Dyke	Fragmental Udic Haplustalfs	SiC, Cl
G	Granitic Pediment Plain		
G1	Crest		
G11	Cultivated	Loamy Skeletal Udic Haplustalfs	Ls, Gr, Sl
		Fine Loamy Udic Haplustalfs	
G12	Waste land	Loamy Skeletal Udic Haplustalfs	Ls, Sl
		Fine Loamy Udic Haplustalfs.	
G2	Side Slope		
G21	Terraced Cultivation	Fine Loamy Udic Haplustalfs	Sl, Cl
		Loamy Skeletal Udic Haplustalfs	
G22	Unterraced Cultivation	Loamy Skeletal Udic Haplustalfs	Gr, Sl
		Loamy Fine Udic Haplustalfs	
G23	Waste Land	Loamy Skeletal Udic Haplustalfs	Sl, Gr,S,L

Depth in cm.	Per cent Slope	Rockiness Class	Erosion	Hectare	Per cent
5	6	7	8	9	10
50-100	3-5	R4	Moderate	-------	--------
50-100	3-5	R4	Moderate to Severe	-------	--------
25-50	15-33	R4	Moderate	-------	--------
25-50	15-33	R4	Moderate to Severe	2,506	3.14
50-100	3-5	R2	Moderate to	1,060	1.33
50-100	3-5	R2	Slight to Moderate	689	0.86
Above 100	1-3	R2	None to Slight	------	--------
Above 100	1-3	R2	None to Slight	302	0.38
25-50	10-33	R2	Severe	597	0.75
Above 100	15-33	R4	Severe	196	0.25
50-100	1-3	R0	Moderate	122	0.15
50-100	1-3	R1	Moderate to Severe	453	0.57
50-100	3-5	R0	None to Slight	20,384	25.60
50-100	3-5	R0	None to Slight	6,041	7.58
25-50	3-5	R1	Moderate to Severe	10,893	13.68

1	2	3	4
G3	Valley		
G31	Partially Cultivated	Fine Aquic Ustochrepts	Cl, SiCl
G32	Cultivated	Fine Loamy Aeric Haplaquepts	Cl, Scl
		Fine Loamy Fluventic Ustochrepts	
B	Buried Pediment		
B1	Plain Cultivation		
B11	Large Parcelling	Fine Typic Chromusterts	Cl, SiCl
		Fine Vertic Ustochrepts	
B12	Medium Parcelling	Fine Vertic Ustochrepts	Scl, SiCl
		Fine Loamy Typic Ustochrepts	
B13	Small Parcelling	Fine Vertic Ustochrepts	Scl, SiCl
B2	Wide Floor Valley	Fine Vertic Haplaquepts	Scl, C
	Paddy Cultivation	Fine Loamy Typic Haplaquepts	
A	Alluvial Plain		
A1	Gently Sloping		
A11	Large Parcelling	Fine Typic Chromusterts	C
A12	Small to Medium Parcelling	Fine Typic Chromusters	C
A2	Nearly Level to Very Gently Sloping		
A21	Large Parcelling	Fine Typic Chromusterts	C
A22	Small Parcelling	Fine Typic Chromusterts	Scl, SiCl
A3	Sloping Land of Interfluve	Fine Vertic Ustochrepts	Sl, Scl
L	Limestone Landscape		
L1	Cultivated	Fine Typic Chromusterts	C
		Fine Typic Pellusterts	
L2	Uncultivated	Fine Typic Chromusterts	C

Temperature Class : Hyperthermic
Mineralogy : Mixed Except Vertisols
Montmorillonitic to Vertisols

Ls : Loamy Sand
GrSl : Gravelley Sandy Loam
Scl : Sandy Clay laom
Cl : Clay loam
SiCl : Silty Clay loam
C : Clay
SiC : Silty Clay

5	6	7	8	9	10
Above 100	1-3	R0	None to Slight	-------	--------
Above 100	1-3	R0	None to Slight	6,462	8.11
Above 100	1-3	----	None to Slight	1,090	1.37
Above 100	1-3	----	None to Slight	1,755	2.20
Above 100	1-3	----	None to Slight	2,342	2.94
Above 100	0-3	----	None to Slight	7,930	9.96
Above 100	3-5	----	None to Slight	----	----
Above 100	0-3	----	None to Slight	----	----
Above 100	0-3	----	None to Slight	3,482	4.37
Above 100	0-3	----	None to Slight	1,639	2.06
50-100	3-5	----	Severe	1,563	1.96
Above 100	1-3	----	None to Slight	1,431	1.79
Above 100	3-5	----	Moderate	1,332	1.67
	Habitation			1,863	2.34
	Tanks			4,996	6.27
	River			354	0.44
	Metalled Road			186	0.23
			TOTAL AREA	79,668	100.00

PHYSIOGRAPHY-CUM-SOIL MAP OF SAMPLE STRIP

DAIGAON TO PIPRI

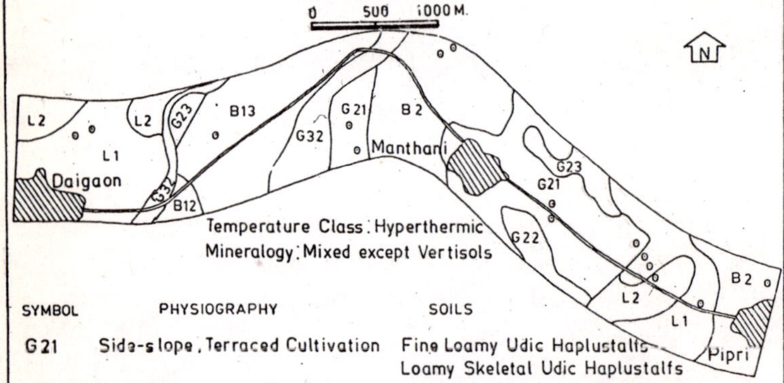

SYMBOL	PHYSIOGRAPHY	SOILS
G21	Side-slope, Terraced Cultivation	Fine Loamy Udic Haplustalfs – Loamy Skeletal Udic Haplustalfs
G22	Side-slope, Unterraced Cultivation	Loamy Skeletal Udic Haplustalfs Fine Loamy Udic Haplustalfs
G23	Side-slope, Waste Land	Loamy Skeletat Udic Haplustalfs
G32	Valley, Pediment Plain	Fine Loamy Aeric Haplaquepts Fine Loamy Fluventic Ustochrepts
B12	Medeum Parcelling, Buried Pediment Plain	Fine Vertic Ustochrepts Fine Loamy Typic Ustochrepts
B13	Small Parcelling, Buried Pediment	Fine Vertic Ustochrepts
B 2	Valley, Buried Pediment	Fine Vertic Haplaquepts Fine Loamy Typic Haplaquepts
L 1	Cultivated, Limestone	Fine Montmorillonitic Typic Chromusterts Fine Montmorillonitic Typic Pellusterts
L 2	Uncultivated, Limestone	Fine Montmorillonitic Typic Chromusterts

G Granitic Pediment Plain
B Buried Pediment
L Limestone Landscape

Metalled Road
Habitation
Soil Boundary (G21)
Observation

PREPARED BY R. S. Yadava

Fig. 9

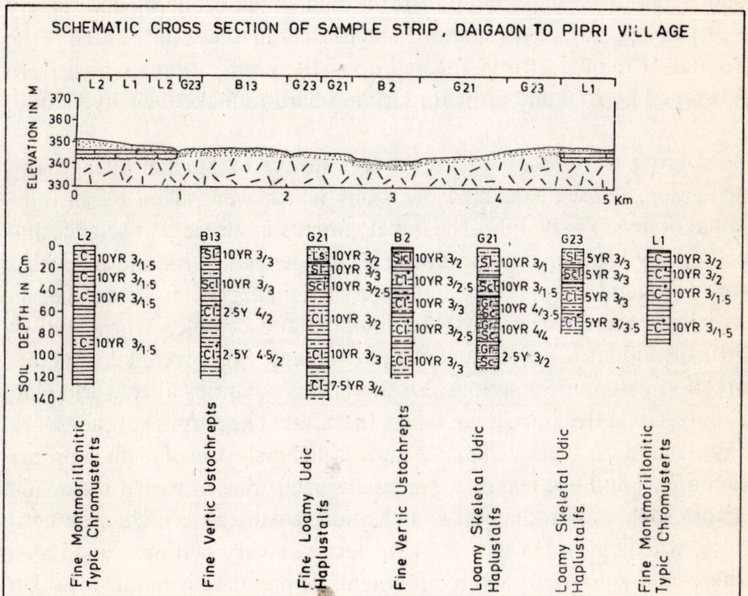
SCHEMATIC CROSS SECTION OF SAMPLE STRIP, DAIGAON TO PIPRI VILLAGE
L2 L1 L2 G23 B13 G23 G21 B2 G21 G23 L1
ELEVATION IN M
370
360
350
340
330
0 1 2 3 4 5 Km
SOIL DEPTH IN Cm
0
20
40
60
80
100
120
140
L2
B13
G21
B2
G21
G23
L1
Fine Montmorillonitic Typic Chromusterts
Fine Vertic Ustochrepts
Fine Loamy Udic Haplustalfs
Fine Vertic Ustochrepts
Loamy Skeletal Udic Haplustalfs
Loamy Skeletal Udic Haplustalfs
Fine Montmorillonitic Typic Chromusterts

Fig. 10

CHAPTER - 4

LAND EVALUATION

By virtue of his scientific pursuits and perception, Man has been able to alter, modify, shape and transform the habitat, economy and society of any area/group of population. The present land use of the area is the product of human efforts enacted upon the earth surface which were developed by trial and error for striving hard to make their livelihood.

During the past, land uses were subject to gradual change and evaluation, consequent upon decisions which were taken by an individual or group of people. The developments in the field of science and technology have made Man aware of the genesis, form and processes leading to land use planning and watershed management. It becomes an essential imperative in densely populated and complex world, full of intrinsic and intricately associated phenomena. Steps were taken in this direction with utmost seriousness in various countries after world war, namely, Great Britain, China, Japan, India, etc. The purpose of such work was to make stock of available resources and make use of them in proper direction. Land use changes, sometime occur due to varied incidental factors such as introduction of irrigation, chemical fertilizers, insecticides, weedicides, HYVs, etc. This becomes very pertinent in an area where changing needs and requirements of population and its resultant pressure involve competitive use of the same piece of land.

The most important function of land use planning is to guide on land use in such a way that the resources of environment are put to the most profitable use in long term perspectives with least adverse impact in nature. The land use planning must be based upon proper understanding of natural environment and the kind of land use. The function of land evaluation is to bring about such understanding and to present planner with a better substitute by comparing the most promising kind of land use.

Land evaluation mainly concerns with the assessment of land performance when used for specific purpose. It involves execution and interpretation of basic survey of climate, soil, vegetation and other attributes of land in terms of their requirements for alternative types of land use within physical, economic, social and cultural context of the area, taken into consideration. Here, an attempt has been made to evaluate the land units of the Nizam Sagar Command, keeping in view the important land qualities and land characteristics of the area which are very relevant. Before going into its complicacy, it is essential to define some specific terms used for such an exercise.

LAND UTILIZATION TYPE(LUT):

A land Utilization Type is a kind of land use which is defined in a degree of detail, greater than that of a major kind of land use. The land use planning and management cannot take into consideration a model different altogether undermining the prevalent practices of land use. Thus, emphasis has been led to make a judicious appraisal of current land use pattern and careful management for fuller utilization of land resources.

LAND UTILIZATION TYPE IN THE PROJECT AREA

The area under study encompasses different types of land formations. Most of the mapping units are fertile and put under cultivation excepting hills and part of pediment plain which reveal compound land utilization type. Various kinds of uses occur in different seasons of year in specific manner (Crop rotation) : paddy-fallow; maize-turmeric-lentil; paddy-paddy; fallow-sugarcane; paddy-wheat; jowar-turmeric-lentil; jowar-groundnut; etc. are popularly found in the study area. The required infrastructure is available for paddy (including summer paddy), sugarcane, maize, jowar, wheat, turmeric, lentil etc. which cover more than ninety per cent of the total cropped area. Turmeric, sugarcane and chillies are the important cash crops which occupy significant proportion of the cropped area. The dominant land utilization types considered in this area are mentioned below :

1. Land Utilization Type I : Paddy ;
2. Land Utilization Type II : Sugarcane ;
3. Land Utilization Type III : Jowar-Maize-Millets ; and
4. Land Utilization Type IV : Turmeric and Caster

LAND QUALITIES FOR LAND CHARACTERISTICS

There are a number of complex attributes of land which act in distinct manner and influence suitability of land for specific type of land utilization. There are a large number of land qualities and land characteristics but only attributes relevant to the study area, have been taken into consideration, namely :

Sl. No.	Land Quality
1.	Soil Texture
2.	Drainage
3.	Inherent fertility
4.	pH
5.	Relief
6.	Slope
7.	Risk of flooding
8.	Organic matter
9.	Soil depth
10.	Erosion Hazard

CRITERIA AND RATING OF LAND QUALITIES

The diagnostic criteria exert influence upon the input and also serve as a basis for assessment of land suitability of given area for that specific land utilization type. Criteria made for different suitability classes for paddy, sugarcane, jowar-maize-millets etc. and, turmeric and caster are given vide in Table 9.

LAND SUITABILITY CLASSIFICATION

Land suitability may be termed as the fitness of a given type of land for defined land utilization type. In this classification, land has been considered in its present condition. The FAO Framework for Land Evaluation has the same structure as in other interpretative classification. Four categories of suitability have been recognized in descending order, i.e. order, classes, sub-classes and units reflecting the kind of suitability within order and reflecting kind of limitation within classes.

STRUCTURE OF SUITABILITY CLASSIFICATION

i) Land suitability orders : reflecting kinds of suitability.

ii) Land suitability classes : reflecting degree of suitability within orders.

iii) Land suitability sub-classes : reflecting kinds of limitations or main kinds of improvement measures required within classes.

iv) Land suitability units : reflecting minor differences in required management within sub-classes.

The present study encompasses land suitability classification upto sub-class level due to certain limitations, viz. kind of survey, scale of map, etc. The categories of classification for Land Evaluation are given in Table 9.

Table 9
Categories of Classification for Land Evaluation

Orders of suitability	Classes of suitability	Sub-classes (kind of limitations)
S-Suitable	Sl - Highly suitable	D- drainage
		d - soil depth
		e - erosion
	S2 - Moderately suitable	f - CEC
		p - pH
	S3 - Marginally suitable	o - Organic matter
		r - relief
		s - slope
		t - texture
		w - risk of flooding
N - Not suitable	No class	

SUITABILITY CATEGORIES AND ITS LIMITATIONS

The mapping units of the area were rated for different LUTs by comparing its land qualities with requisite land qualities for respective LUTs. Such an exercise was done for all the LUTs seprately as the very unit which is highly suitable for one LUT, may be not suitable for other LUTs. Besides this, a few units may be highly suitable for more than one LUT and moderately suitable for third and marginally suitable for fourth LUT.

This exercise on the one hand helps to understand the validity of current farming practices evolved on the basis of trial and error on one side and concludes with a better alternate land utilization type based upon the inherent quality of land units on the other. Soil suitability for various land units based on their qualities for four prevalent LUTs has been worked out separately and summarised in Table 10, 11, 12 and 13. Emphasis has been given to highlight common limitations of land units for each category of land utilization type.

Highly Suitable (S1)

The mapping units which did not have any significant limitation for its sustained use for given land utilization type, are classed as highly suitable for that particular LUT.

Moderately Suitable (S2)

Common limitations marked in various mapping units classed under this category are soil depth, relief and risk of flooding which are permanent in nature and cannot be improved for its sustained use. Whereas limitations of fertility, pH, organic matter and erosion hazard are temporary and may be improved by adding variety manure, fertilizer, chemical amendment on the one hand and by adopting suitable soil and water conservation measures as embraced in integrated watershed management. These soils may become highly suitable for specific LUTs by adopting suitable measures for improvement in such limitations.

Marginally Suitable (S3)

The permanent limitations of this category are depth of soil, relief and risk of flooding which cannot be improved and used for sustained utilization of land for a defined land use. The improvable limitations are fertility, organic matter, erosion hazard which may be ameliorated after adding a few chemicals / and adopting suitable measures of soil conservation. Accordingly, it may become moderate to highly suitable for cultivation in days to come when pressure of population will substantially increase in the area.

Not Suitable (N)

The soils of this category are not suitable for sustained use of any land utilization type because of its permanent limitations, namely, relief, soil depth, risk of flooding and steepness of slope. These cannot be improved even after incur of significant amount of labour, mechanical power, chemicals, etc. At the same time, cost of alternation shall be such a higher side that individual farmers cannot afford it. Such areas need to be left as such so that natural regeneration of suitable species of flora and fauna may take place and improve ecosystem. Its feasibility may be worked out for growing grapes under careful management practices without any side effect on the environment.

Table 10
Criteria And Rating Of Land Qualities For Paddy (LUT - I)

Sl. No.	Land Quality	Suitability Class S1	S2	S3	Not Suitable N
1.	Texture	Fine loamy	Coarse loamy	Sandy	Fragmental
2.	Drainage	Imperfectly drained and poorly drained	Moderate to well drained	Some-what excessi-vely drained	Excessively drained
3.	Inherent fertility (CEC)	High	Medium	Low	Very low
4.	pH	6-5-7.5	7.5-8.5 6.0-6.5	8.5-9.5 5.0-6.0	Above 9.5 Below 5.0
5.	Relief	Flat and subnormal	Concave	Normal	Excessive
6.	Slope	Level	Gently sloping	Undula-ting	Moderately sloping
7.	Organic carbon	Medium to high	Low to medium	Low	Very Low
8.	Soil depth	Very deep to deep	Moderately deep	Shallow	Very Shallow
9.	Erosion hazard	None to slight	Moderate	Severe	Very Severe
10.	Risk of flooding	Low	Medium	High (short duration)	Very High (long duration)

Table 11
Criteria And Rating Of Land Qualities For Sugarcane (LUT - II)

Sl. No.	Land Quality	Suitability Class			Not Suitable
		S1	S2	S3	N
1.	Texture	Clay loam and heavy loam	Ligh loam, loam, clay	Sandy loam	Sandy, fragmental
2.	Drainage	Moderately well-drained	Imperfectly and well-drained	Poorly drained	Excessively drained
3.	Inherent fertility (CEC meg/ 100 g soil)	Very High Above 20	High 15-20	Medium 10-15	Low Below 10
4.	pH	6-5-7.5	7.5-8.5 6.0-6.5	8.5-9.5 5.5-6.0	Above 9.5 Below 6.5
5.	Relief	Flat to subnormal	Normal	Concave	Excessive
6.	Organic carbon	High to medium	Medium to low	Low	Very low
7.	Soil depth	Very deep	Deep	Moderately deep	Shallow
8.	Slope	Flat to nearly level	Gently sloping	Moderately sloping	Strongly sloping
9.	Erosion hazard	None to slight	Moderate	Severe	Very severe
10.	Risk of flooding	Low	Moderate	High and short duration	Very high and long duration

Table 12
Criteria And Rating Of Land Qualities For Jowar, Maize and Millets (LUT - III)

Sl. No.	Land Quality	Suitability Class S1	S2	S3	Not Suitable N
1.	Texture	Loamy	Coarse loamy and clayey	Sandy	Fragmental
2.	Drainage	Well drained	Moderately well drained	Imper-fectly drained	Excessive and poorly drained
3.	Inherent fertility (CEC)	High to medium	Medium	Low	Very low
4.	pH	6-5-7.5	7.5-8.5 6.0-6.5	8.5-9.5 5.0-6.5	Above 9.5 Below 5.0
5.	Relief	Normal	Subnormal	Concave	Excessive
6.	Slope	Level	Gently sloping	Undu-lating	Strongly sloping
7.	Organic carbon	High to medium	Medium	Low	Very low
8.	Soil depth	Very deep to deep	Moderately deep	Shallow	Very shallow
9.	Erosion hazards	None to slight	Moderate	Severe	Very Severe
10.	Risk of flooding	Low	Medium	High	Very High

Table 13
Criteria And Rating Of Land Qualities For Turmeric And Caster (LUT - IV)

Sl. No.	Land Quality	Suitability Class S1	S2	S3	Not Suitable N
1.	Texture	Loamy	Clayey and light loamy	Sandy	Fragmental
2.	Drainage	Well drained	Moderately well drained	Imper-fectly drained	Poor and excessively drained
3.	Inherent fertility (CEC)	High	Medium	Low	Very low
4.	pH	6-5-7.5	7.5-8.5 6.0-6.5	8.0-8.5 5.0-6.0	8.5 & Above Below 5.0
5.	Relief	Normal	Subnormal	Concave	Excessive
6.	Slope	Level	Gently sloping	Undu-lating	Moderately sloping
7.	Organic carbon	High to medium	Medium to low	Low	Very low
8.	Soil depth	Very deep to deep	Deep	Mode-rately deep	Shallow
9.	Erosion hazards	None to slight	Moderate	Severe	Very Severe
10.	Risk of flooding	Very Low	Low	Mode-rate	High

LAND SUITABILITY FOR PADDY CULTIVATION (LUT - I)

All the land units of the study area have been grouped under various land suitability order reflecting kind of suitability, land suitability class reflecting degree of suitability within order and land suitability sub-class-reflecting kinds of limitations or kinds of improvement measures required within the class on the basis of prevalent land qualities rates for this land utilization type as compared with required ones in semi arid type of climate. (Table 14)

Highly Suitable Sub-Class :

After an exhaustive exercise for land evaluation, it has been found that A22, B13 and B2 units are highly suitable for paddy cultivation. These units exhibit highly suitable all the land qualities considered for this LUT namely fine loamy texture, imperfectly and poorly drained condition, high fertility, 6.5-7.5 pH, flat to sub-normal relief, level land, medium to high organic matter, very deep to deep soil, none to slight erosion and low risk of flooding. It occupies an area of 11,911 ha and consists 14.95% area of the survey. (Table 15)

Moderately Suitable Sub-Class :

Unit H42, G21, G22, G32, B11, B12, A11, A12, L1 and L2 are burdened to some extent due to a few limitations such as coarse loamy texture, moderate to well drained condition, medium inherent fertility, 7.5-8.5 and 6.0-6.5 pH, concave relief, gently sloping, low to medium organic matter, moderately deep soil, moderate erosion and medium risk of flooding, and rendered moderately suitable for this LUT. These units are spread over 42,279 ha which comprise 53.07% of the area.

Presently, H42 unit is under rainfed paddy cultivation, under limitations of concave relief and medium risk of flooding. The adoption of soil conservation measures may make the unit highly suitable for paddy cultivation.

Unit G21 is characterised with limitations of moderate to well drained and medium inherent fertility whereas G22 with well drained to moderately well drained and qualify to this suitability sub-class.

G32 unit is burdened with concave relief and medium risk of flood and may be allowed to remain under paddy cultivation with desirable measures of soil conservation.

Table 14
Land Qualities Of Various Mapping Units Of Lower Nizam Sagar Command

Sl. No.	Mapping unit	Texture	Drainage	Inherent fertility CEC	pH
1	2	3	4	5	6
1.	H11	Sandy loam to clay loam	Well to somwhat excessively drained	Medium	6.7
2.	H12	Sandy loam to sandy clay loam	Medium	Medium	6.8
3.	H21	Sandy loam to sandy clay loam	Excessively drained	Medium	6.6
4.	H22	Sandy loam to clay loam	Excessively drained	Medium	6.8
5.	H31	Sandy loam to sandy clay loam	Well to excessi-vely drained	Medium	6.8
6.	H32	Sandy loam to clay loam	Well drained	Medium	6.8
7.	H41	Silt loam to silty clay loam	Imperfect to poorly drained	High	7.6
8.	H42	Silt loam to clay	Poorly drained	Very high	7.8
9.	H5	Sandy loam to sandy clay loam	Excessively drained	Medium	6.8
10.	H6	Loam to clay	Excessively drained	Very High	7.9
11.	G11	Loamy sand to clay loam	Well drained	Very High	7.2
12.	G12	Sandy loam to clay loam	Well drained	Medium	6.8
13.	G21	Sandy loam to silty clay loam	Moderate to well drained	Medium to high	7.1
14.	G22	Loam to clay loam	Well drained to moderately well drained	Medium to High	7.3
15.	G23	Sandy loam to sandy clay loam	Well drained	Medium to Low	6.8
16.	G31	Silt loam to silty clay loam	Poorly drained	Medium to High	7.6

Relief	Slope	Organic matter (%)	Soil depth	Erosion hazard	Risk of flooding
7	8	9	10	11	12
Normal	Gently sloping	High	Deep	Moderate	None
Normal	Gently sloping	High	Deep	Moderate to Severe	None
Excessive	Strong to steeply sloping	High	Moderately deep	Moderate	None
Excessive	Strongly steep	Medium	Moderately deep	Severe	None
Normal to excessive	Gently sloping	Medium	Deep to very deep	Moderate to severe	None
Normal	Gently sloping	Medium	Deep to very deep	Slight to Moderate	None
Concave	Nearly level	High	Very deep	None	High
Concave	Nearly level	High	Very deep	None	Moderate to high
Excessive	Moderately steep to steep	Low	Moderately deep	Severe	None
Excessive	Moderately steep to steep	High	Very deep	Severe	None
Sub-normal	Nearly level	Medium	Deep to very deep	Slight to Moderate	None
Sub-normal	Nearly level	High	Deep to very deep	Moderate	None
Normal	Nearly level to gently sloping	High	Deep to very deep	Slight	None
Sub-normal	Nearly level	High	Deep to very deep	Slight	None
Normal	Gently undulating	Low	Very deep (skeletal)	Severe	None
Concave	Nearly level	High	Very deep	None	Moderate to High

1	2	3	4	5	6
17.	G32	Silt loam to silty clay loam	Poorly drained to moderately drained	High	7.8
18.	B11	Clay loam to clay	Moderately well drained	High	7.8
19.	B12	Clay loam to clay	Imperfectly to moderately well drained	High	8.0
20.	B13	Clay loam to clay	Moderately well drained & imperfectly drained	High	8.0
21.	B2	Silt loam to clay	Poorly drained	High	8.4
22.	A11	Clay	Moderately well drained	High	7.7
23.	A12	Clay	Moderately well drained	High	7.8
24.	A21	Clay loam to clay	Moderately well to imperfectly drained	High	8.0
25.	A22	Clay loam to clay	Moderately well to imperfectly drained	High	8.2
26.	A3	Sandy loam to clay	Well drained	Moderate	7.8
27.	L1	Clay loam to clay	Moderately well drained	High	8.4
28.	L2	Clay loam to clay	Moderately well drained	High	8.4

7	8	9	10	11	12
Concave	Nearly level	High	Very deep	None	Moderate
Sub-normal	Nearly level	High	Very deep	Slight	None
Sub-normal	Nearly level	High	Very deep	Slight	None
Sub-normal	Gently sloping	High	Very deep	Slight to Moderate	Moderate to Low
Concave	Nearly level	High	Very deep	None	Moderate to High (short duration)
Normal	Gently sloping	High	Very deep	Slight	Low
Normal	Gently sloping	High	Very deep	Slight	Moderate to low
Sub-normal	Nearly level	High	Very deep	Slight	Moderate to low
Sub-normal	Nearly level	High	Very deep	Slight	Moderate to low
Sub-normal	Gently sloping	Medium	Very deep	Severe	Low
Normal to subnormal	Nearly level to gently sloping	Medium	Very deep	Slight	Low
Normal to subnormal	Gently sloping	Medium	Very deep	Moderate	Low

Table 15
Rating Of Land Qualities Of Various Mapping Units For Paddy (LUT - I)

Sl. No.	Mapping unit	Texture	Drainage	Inherent fertility (CEC)	pH	Relief	Slope	Organic matter	Soil depth	Erosion hazard	Risk of flooding	Suitability Classification: Order	Class	Sub - Class
1	2	3	4	5	6	7	8	9	10	11	12	13	14	15
1.	H11*	S2	N	S2	S1	N	S2	S1	S1	S2	S1	N	N	NDr
2.	H12*	S2	N	S2	S1	S2	S2	S1	S1	S3	S1	N	N	NDr
3.	H21*	S2	N	S2	S1	N	N	S1	S2	S3	S1	N	N	NDrs
4.	H22	S2	N	S2	S1	N	N	S2	S2	S3	S1	N	N	NDrs
5.	H31	S1	N	S2	S1	N	S3	S1	S1	S3	S1	N	N	NDrs
6.	H32	S2	S2	S2	S1	S2	S2	S1	S1	S3	S1	N	N	NDrs
7.	H41*	S1	S3	S1	S1	S3	S1	S1	S1	S1	S1	S	S2	S2Dr
8.	H42	S1	S1	S1	S2	S2	S1	S1	S1	S1	S2	S	S3	S3Dr
9.	H5	S2	N	S2	S1	N	N	S1	S1	S3	S2	S	S2	S2rw
10.	H6	N	N	S1	S1	N	N	S1	S1	S3	S1	N	N	NDrs
11.	G11	S2	S2	S2	S1	S1	S1	S2	S1	S1	S1	N	N	NDrs
12.	G12	S2	S2	S2	S1	S1	S1	S2	S1	S2	S1	S	S2	S2Df
13.	G21	S1	S2	S2	S1	S1	S1	S1	S1	S1	S1	S	S2	S2Dfe
14.	G22	S1	S2	S1	S1	S1	S1	S1	S1	S1	S1	S	S2	S2Df
15.	G23	S2	S2	S2	S1	S3	S2	S2	S2	S1	S3	S	S2	S2D
16.	G31*	S3	S1	S1	S1	S2	S1	S1	S1	S1	S2	S	S3	S3re
17.	G32	S1	S1	S1	S1	S2	S1	S1	S1	S1	S2	S	S3	S3trw
18.	B11	S1	S2	S1	S2	S1	S1	S1	S1	S1	S1	S	S2	S2rw
19.	B12	S1	S2	S1	S2	S1	S1	S1	S1	S1	S1	S	S2	S2Dp
20.	B13	S1	S1	S1	S1	S1	S1	S1	S1	S1	S1	S	S2	S2Dp
21.	B2	S1	S1	S1	S1	S1	S1	S1	S1	S1	S1	S	S1	S1
22.	A11*	S2	S2	S1	S2	S2	S2	S1	S1	S1	S1	S	S1	S1
23.	A12*	S2	S2	S1	S2	S2	S2	S1	S1	S1	S1	S	S2	S2Drp
24.	A21	S1	S2	S1	S1	S1	S1	S1	S1	S1	S1	S	S2	S2Dpr
25.	A22	S1	S1	S1	S1	S1	S1	S1	S1	S1	S2	S	S2	S2D
26.	A3	S2	S2	S2	S1	S2	S2	S2	S1	S3	S1	S	S1	S2
27.	I1	S2	S2	S1	S1	S1	S1	S2	S1	S1	S1	S	S3	S3e
28.	L2	S2	S2	S1	S1	S1	S1	S2	S1	S1	S1	S	S2	S2D

* The mapping Units - H11, H12, H21, H41, H31, A11 and A12 are observed and mapped in adjoining Bodhan area.

B11 and B12 units are moderately suitable for its current LUT wherein limitations of imperfectly to moderately well drained and 7.5-8.0 pH are prevalent. But these units are highly suitable for sugarcane cultivation. These units may well be used for LUT-II for its commercial value as sugar mills are situated very close to it.

A11 unit is affected due to its clay texture, moderately well drained, 7.2 pH, normal relief and gently sloping, and B12 unit with clay loam to clay texture and 8.0 pH. Thus, these two units are rendered moderately suitable for this LUT. In due course of time, these units may be used for paddy and sugarcane cultivation.

Unit L1 and L2 are moderately suitable for paddy and sugarcane cultivation. Here, clay loam to clay texture of soil, moderately well drained, drainage and medium organic matter are major limitations. Had there been sufficient water available for irrigation as anticipated in development of Nizamsagar, it would have become highly suitable for paddy cultivation ?

Marginally Suitable Sub-Class

The land units which exhibit a few severe limitations like sandy texture, somewhat excessively drained condition, low inherent fertility, 8.5-9.5 and 5.0-6.0 pH, normal relief, undulating slope, shallow depth of soil, severe erosion and high (short duration) flood. In addition to this many limitations of land qualities of lesser degree for this LUT have also been encountered. These have rendered H41, G23, G31, and A3 units marginally suitable for paddy cultivation (Table 15). These units are spread over 13,145 ha and constitute 16.50% of its total area.

Unit H41 is marginally suitable for this land utilization type but due to its submergence for storage of water for irrigation, it cannot be used for cultivation.

H41 is marginally suitable for paddy cultivation due to concave relief and imperfect to poorly drained condition. This LUT may be continued subject to limitations of lesser degree inherent fertility, nearly level, low organic matter and high risk of flood. Second category of problems are not much significant during the year of normal amount of rainfall during the monsoon period. Green manures and decomposed organic matter may be added to increase per ha yield and efficiency of land utilization.

Unit G23 reflects normal relief and severe erosion hazard. Besides these, low inherent fertility, organic matter and soil depth pose constraints of lesser degree. This unit remains unterraced but it is put under cultivation. This unit may be used for growing sugarcane and in turn will add to the prosperity of people.

G31 unit currently remains waste land in which gravelly sandy loam texture poses severe limitations. In addition to this, concave relief and moderate to high risk of flooding are other significant constraints. Provision of moisture retention and suitable terracing may enhance its suitability for this LUT. Whereas, A3 unit which is characterized with sloping land of interfluves, happens to be marginally suitable for this land utilization type (Fig. 11, 12 and 13).

Not Suitable For Paddy Cultivation :

Besides these H11, H12, H22, H5, H6, G11, G12 units are not suitable for any of the land utilization type taken into consideration. All the units of hill landscape are burdened with some sort of permanent constraints, namely, excessive drainage condition, excessive relief and moderately sloping. Due to lack of sufficient soil moisture, water for irrigation, coarse texture, CEC, etc., it is not suitable for all those land utilization types. These may successfully be brought under grape cultivation or safely put under agro-forestry practices.

LAND SUITABILITY FOR SUGARCANE CULTIVATION (LUT-II)

Land units were appraised for various land suitability order- reflecting kind of suitability, land suitability class-reflecting degree of suitability within the order and land suitability sub-class-reflecting kinds of limitations or kinds of improvement measures within the class on the basis of comparison of its qualities with required ones for sugarcane cultivation in the command area. (Table 16).

Highly Suitable Sub-Class :

The preponderance of sugarcane cultivation in this part and upper command area, sought the need for land evaluation for this land utilization type. Mapping units characterized with clay loam and heavy loam texture, moderately well drained, very high CEC, 6.5-7.5 pH, flat to sub-normal relief, high to medium organic matter, very deep soil, flat to nearly level, none to slight erosion and low risk of flooding are classed as highly suitable for LUT-II. The unit B11 and B12 exhibit all these land qualities and are highly suitable for this LUT. These units are spread over 2,845 ha and constitute 3.57% of the survey area. These units are free from

SOIL MAP OF PART OF NIZAMABAD DISTRICT, A.P.

Lower Nizamsagar Command, Binola-Navipet Area

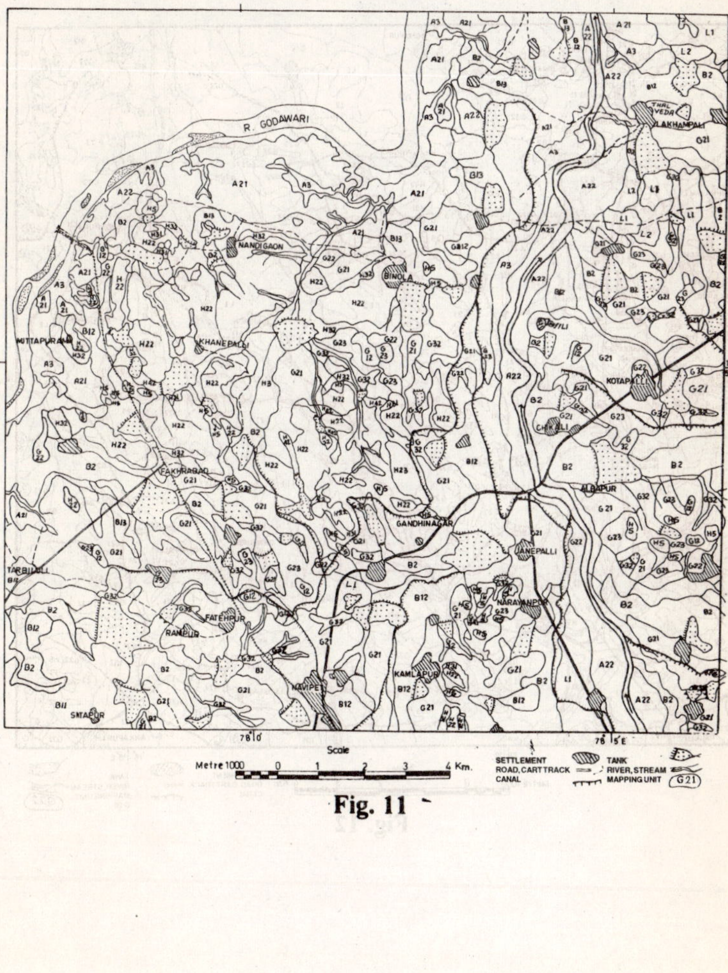

Fig. 11

SOIL MAP OF PART OF NIZAMABAD DISTRICT, A.P.

Lower Nizamsagar Command, Navipet-Anksapuram Area

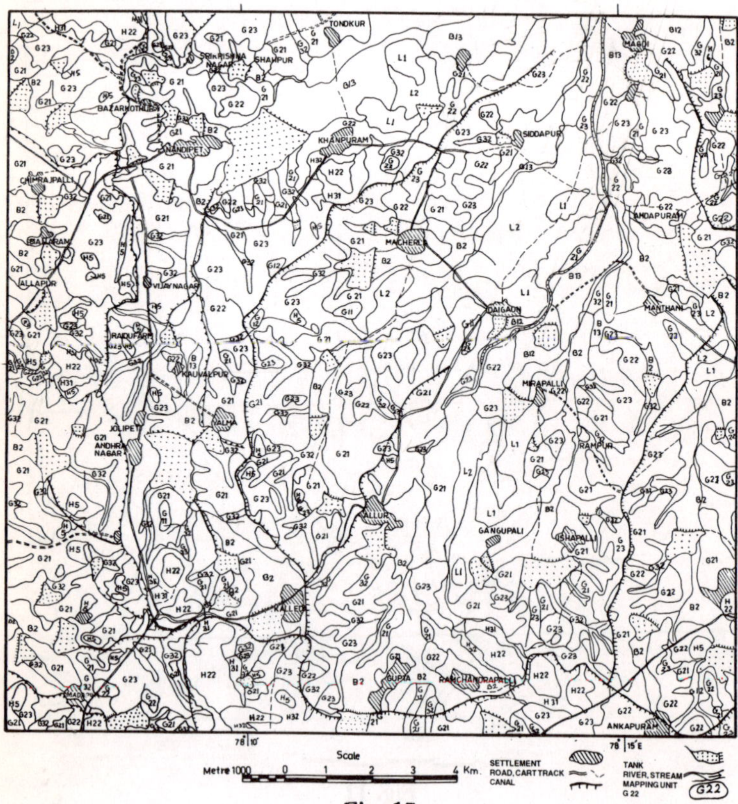

Fig. 12

SOIL MAP OF PART OF NIZAMABAD DISTRICT, A.P.

Lower Nizamsagar Command, Balkonda - Yelpur Area

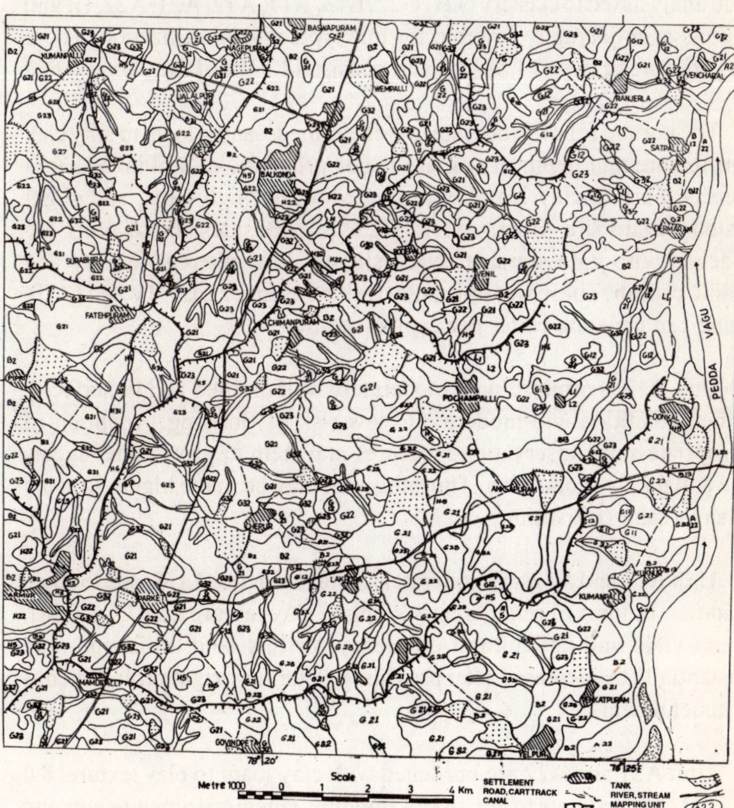

Fig. 13

any permanent constraint, posing any problem to this land utilization type.

Moderately Suitable Sub-Class :

The mapping unit with a few moderate limitations in form of light loam, loamy clay texture, high CEC, 7.5-8.5 and 6.0-6.5 pH, medium to low organic matter, deep soil, gently sloping, moderate erosion and moderate flood happen to be classed as moderately suitable for this LUT. Vivid analysis led to classify G21, G22, B13, A11, A12, A21, A22, L1 and L2 units as moderately suitable for this type of land utilization. These units are spread over 34,884 ha and consist 44.34% of the survey area.

The normal relief, moderate to well drained, sandy loam to silty clay loam texture and depth of soil are major problems of soil in G21 unit whereas loam to clay loam texture, medium to high CEC, and slight erosion hazard are prominent in G22 unit. Besides, these units are burdened with unfavourable relief, slightly low inherent fertility and organic carbon. These two units may well be used for LUT-IV and LUT-III and can generate better per capita income.

Unit B13 is affected with moderately well drained and imperfectly drained condition and moderate to low risk of flooding. Adoption of appropriate soil conservation measures may help to grow better sugarcane and paddy production. Thus, this unit may be made highly suitable for sugarcane cultivation.

Unit A11 and A12 are moderately suitable for this LUT, due to some limitation in form of relief, gently sloping, clay texture and 7.5-8.0 pH. Per ha yield and agricultural efficiency for this land utilization may substantially be increased after providing water for irrigation and chemical amendments.

Unit A21 and A22 are burdened with clay loam to clay texture, 8.0 - 8.5 pH and moderate to low risk of flooding. Chemical amendments and harvesting of run-off at suitable sites may be able to improve production of this LUT of the landscape.

L1 and L2 are burdened with heavy soil in form of clay loam to clay texture and 8.4 pH. In addition to these qualities, L2 is characterised with gently sloping and moderate erosion. Accordingly, these units are moderately suitable for sugarcane cultivation. Insufficient water available for irrigation from Nizamsagar command could not yield desired result for increase in its agricultural productivity.

Table 16
Rating Of Land Qualities Of Various Mapping Units for Sugarcane Cultivation (LUT - II)

Sl. No.	Mapping unit	Texture	Drainage	Inherent fertility(CEC)	pH	Relief	Slope
1	2	3	4	5	6	7	8
1.	H11*	S2	N	S2	S1	N	S2
2.	H12*	S2	N	S2	S1	N	S2
3.	H21*	S2	N	S2	S1	N	N
4.	H22	S2	N	S2	S1	N	N
5.	H31	S2	N	S3	S2	N	S2
6.	H32	S2	S3	S3	S2	S2	S2
7.	H41*	S1	S3	S1	S2	S3	S1
8.	H42	S1	S3	S1	S2	S3	S1
9.	H5	S2	N	S2	S1	N	N
10.	H6	S1	N	S1	S2	N	N
11.	G11	S2	S2	S2	S1	S2	S1
12.	G12	S2	S2	S2	S1	S2	S1
13.	G21	S2	S2	S2	S1	S2	S1
14.	G22	S2	S1	S2	S1	S1	S1
15.	G23	S2	S2	S3	S1	S2	S2
16.	G31*	S1	S3	S1	S2	S3	S1
17.	G32	S1	S3	S1	S2	S3	S1
18.	B11	S1	S1	S1	S2	S1	S1
19.	B12	S1	S1	S1	S1	S1	S1
20.	B13	S1	S2	S1	S1	S1	S1
21.	B2	S2	S3	S1	S2	S3	S1
22.	A11*	S2	S1	S1	S2	S2	S2
23.	A12*	S2	S1	S1	S2	S2	S2
24.	A21	S2	S1	S1	S2	S1	S1
25.	A22	S2	S1	S1	S2	S1	S1
26.	A3	S2	S2	S2	S2	S2	S2
27.	L1	S2	S1	S1	S2	S1	S1
28.	L2	S2	S1	S1	S2	S1	S2

Contd. p- 66

Organic matter	Soil depth	Erosion hazard	Risk of flooding	Suitability Classification Order	Class	Sub-Class
9	10	11	12	13	14	15
S1	S2	S3	S1	N	N	NDr
S1	S2	S3	S1	N	N	NDr
S1	S3	S3	S1	N	N	NDrs
S2	S3	S3	S1	N	N	NDrs
S2	S1	S3	S1	N	N	NDr
S2	S1	S1	S1	S	S3	S3Df
S2	S1	S1	N	N	N	ND
S2	S1	S1	N	N	N	ND
S2	S2	S3	S1	N	N	NDrs
S2	S1	N	S1	N	N	NDrs
S2	S2	S1	S1	S	S2	S2Ddf
S1	S2	S2	S1	S	S2	S2Ddfe
S1	S2	S1	S1	S	S2	S2Ddf
S1	S2	S1	S1	S	S2	S2Ddf
S3	S3	S3	S1	S	S3	S3Dde
S1	S1	S1	N	N	N	NDw
S1	S1	S1	S3	S	S3	S3Drw
S1	S1	S1	S1	S	S1	S1
S1	S1	S1	S1	S	S1	S1
S1	S1	S1	S2	S	S2	S2Dw
S1	S1	S1	S3	S	S3	S3Drw
S1	S1	S1	S1	S	S2	S2prs
S1	S1	S1	S1	S	S2	S2prs
S1	S1	S1	S2	S	S2	S2pw
S1	S1	S1	S2	S	S2	S2pw
S2	S2	S3	S1	S	S3	S3te
S1	S1	S1	S1	S	S2	S2p
S1	S1	S2	S1	S	S2	S2tpe

* The mapping units H11, H12, H21, H41, G31, A11 and A12 are observed and mapped in adjoining Bodhan area.

Marginally Suitable Sub-Class :

The mapping unit with a few severe limitations in form of sandy loam texture, poorly drained condition, medium CEC, 8.5-9.5 and 5.5 - 6.0 pH, concave relief, low organic matter, moderately deep soil, moderately sloping, severe erosion and high (short duration) flood, and many limitations of lesser degree are responsible for their classification under marginally suitable for LUT-II. It has been noted that H32, G23, B2 and A3 units are marginally suitable for sugarcane cultivation. These units encompass 27,537 ha and constitute 34.56% of the total survey area.

H32 unit is not that congenial for its well drained condition and medium inherent fertility to a marginal extent on one hand and sandy loam to clay texture, 6.8 pH, relief, and medium organic carbon on the other. Thus, relief and drainage are responsible for its marginal suitability for this LUT.

G23 is characterised with medium to low fertility, low organic carbon, deep but with skeletal material and severe erosion hazard. Besides these, moderate suitability of texture, drainage, relief and slope have also been noticed in this unit.

The mapping unit G32 is burdened with severe limitations of drainage and relief, and moderate pH for this LUT. Thus, it is classed as marginally suitable in this respect.

B2 unit suffers from acute poor drainage and concave relief and A3 from unfavourable topography and severe erosion hazard. Adoption of appropriate measures may help in increase of per ha yield and ameliorate land suitability class for this crop land use.

Not Suitable for LUT-II :

Besides these, H11, H12, H21, H22, H31, H41, H42, H5, H6, G11 and G12 units exhibit many permanent limitations for this LUT. Major limitations are relief, drainage, slope and risk of flooding. Being located in semi-arid climatic condition and its low moisture retention, granitic pediments with loamy skeletal Udic Haplustalfs are not suitable for this land utilization type. This unit occupy 4,176 ha and consist of 5.24% of the total area.

LAND SUITABILITY FOR JOWAR, MAIZE & MILLETS (LUT-III)

It is worth to mention that land units not suitable for paddy or sugarcane cultivation are quite suitable for this type of land utilization, as its land qualities are favourable for growing such crops. The summary of suitability categories reveals that 67,608 ha are suitable for growing such crops which constitute 74.86% of the study area. Although, the suitability category of different units considerably varies from unit to unit. Similarly, units with all highly suitable qualities, somewhat, and much more deviating from the required one, are termed highly, moderately and marginally suitable for this land utilization type respectively.

Highly Suitable Sub-Class :

It may be noted that soils with loam texture, well drained, high to medium inherent fertility, 6.5-7.5 pH, normal relief, level land, high to medium content of organic matter, deep to very deep soil, none to slight erosion and low risk of flooding are highly suitable for this land utilization (Table 17). G21 and G22 units of this landscape exhibit similar characteristics, hence classed as highly suitable for this land utilization type. These units are spread over 26,425 ha and constitute 33.17% of the area.

Moderately Suitable Sub-Class :

Mapping units with coarse loamy to clay texture, moderately well drained, medium inherent fertility, 7.5-8.5 and 6.0-6.5 pH, sub-normal relief, gently sloping, medium organic matter, moderately deep soil, moderate erosion and medium risk of flooding reflect moderate suitability of soil for this land utilization type. As a few such land qualities have been found in H32, G11, G12, B11, A11, A12, A21, A22, L1 and L2 units, those are classed moderately suitable for this LUT which jointly occupy 11,993 ha of land and consist 15.04% of the total area.

H32 unit assumes this character by virtue of moderate drainage, inherent fertility and organic matter. The drainage condition is the main problem which comes in the way of the unit to qualify for high suitability whereas other qualities can be improved by human efforts.

G11 unit is burdened with slope constraint whereas G12 with low organic matter and erosion hazard. Chemical amendments and measures of soil conservation may enable G12 unit assume high suitability.

B11, B12, A11 and A12 units remain under this category as drainage, pH and texture are the main problems for these units. A21 and A22 exhibit risk of flooding in addition to the problem related to aforesaid units.

Table 17
Rating Of Land Qualities Of Various Mapping Units for Jowar, Maize and Millets (LUT - III)

Sl. No.	Mapping unit	Texture	Drainage	Inherent fertility(CEC)	pH	Relief	Slope
1	2	3	4	5	6	7	8
1.	H11*	S2	N	S2	S1	N	S2
2.	H12*	S2	N	S2	S1	N	S2
3.	H21*	S2	N	S2	S1	N	N
4.	H22	S2	N	S2	S1	N	N
5.	H31	S1	N	S2	S1	N	S1
6.	H32	S1	S2	S2	S1	S1	S1
7.	H41*	S2	N	S1	S2	S3	S1
8.	H42	S2	N	S1	S2	S3	S1
9.	H5	S1	N	S2	S1	N	N
10.	H6	S2	N	S1	S2	N	N
11.	G11	S1	S1	S1	S1	S1	S2
12.	G12	S1	S1	S1	S1	S1	S1
13.	G21	S1	S1	S1	S1	S1	S1
14.	G22	S1	S1	S1	S1	S1	S1
15.	G23	S2	S1	S2	S1	S1	S2
16.	G31	S2	N	S1	S1	S1	S1
17.	G32	S2	S3	S1	S1	S3	S1
18.	B11	S2	S2	S1	S2	S1	S1
19.	B12	S2	S2	S1	S2	S1	S1
20.	B13	S2	S2	S1	S2	S1	S1
21.	B2	S2	S3	S1	S2	S3	S1
22.	A11*	S2	S2	S1	S2	S1	S1
23.	A12*	S2	S2	S1	S2	S1	S1
24.	A21	S2	S2	S1	S2	S1	S1
25.	A22	S2	S2	S1	S2	S1	S1
26.	A3	S2	S1	S1	S2	S1	S1
27.	L1	S2	S2	S1	S2	S1	S1
28.	L2	S2	S2	S1	S2	S1	S1

Contd. p-70

Organic matter	Soil depth	Erosion hazard	Risk of flooding	Suitability Classification		
				Order	Class	Sub - Class
9	10	11	12	13	14	15
S1	S1	S2	S1	N	N	NDr
S1	S1	S3	S1	N	N	NDr
S1	S2	S2	S1	N	N	NDrs
S1	S2	S3	S1	N	N	NDrs
S2	S1	S3	S1	N	N	NDr
S2	S1	S1	S1	S	S2	S2Dfo
S1	S1	S1	N	N	N	NDw
S1	S1	S1	N	N	N	NDw
S1	S1	S3	S1	N	N	NDrs
S1	S1	S3	S1	N	N	NDrs
S1	S1	S1	S1	S2	S2	S2o
S2	S1	S2	S1	S	S2	S2oe
S1	S1	S1	S1	S	S1	S1
S1	S1	S1	S1	S	S1	S1
S1	S1	S3	S1	S	S3	S3e
S1	S1	S1	N	N	N	ND
S1	S1	S1	S3	S	S3	S3Drw
S1	S1	S1	S1	S	S2	S2Dpt
S1	S1	S1	S1	S	S2	S2Dpt
S1	S1	S1	S3	S	S3	S3w
S1	S1	S1	S3	S	S3	S3Drw
S1	S1	S1	S1	S	S3	S2tDpw
S1	S1	S1	S1	S	S2	S2tDpw
S1	S1	S1	S1	S	S2	S2tDpw
S1	S1	S1	S2	S	S2	S2tDpw
S2	S1	S3	S1	S	S3	S3e
S1	S1	S1	S1	S	S2	S2tDp
S1	S1	S2	S1	S	S2	S2tDpe

* These seven units are observed and mapped in the adjoining Bodhan area.

Unit L1 is affected due to texture, drainage and pH while L2 faces erosion hazard as well. It is remarkable that most of the units bear some character which cannot be modified.

Marginally Suitable Sub-Class :

Unit G23, G32, B13, B2 and A3 are marginally suitable for this land utilization type. These units exhibit either of the land quality defined as marginally suitable for this LUT, namely, sandy loam texture, poor drainage, medium inherent fertility, 8.5-9.5 and 5.5-6.0 pH, concave relief, low organic matter, moderately deep soil profile, moderately sloping, severe erosion hazard, high (short duration) flood and come under this land suitability sub-class. In addition to these, many limitations of land qualities of lesser degree of significance have also been noted. These units are spread over 29,190 ha which consist 26.65% of the area.

G23 is burdened with severe erosion hazard whereas G32 with relief, drainage and risk of flooding.

B13 is severely affected by risk of flooding and B21 is unsuitable due to drainage, relief and risk of flooding which cannot be ameliorated.

Besides these, A3 unit is adversely affected with erosion hazard. These units are well under use of some other land utilization type and able to sustain better as well as lucrative LUT.

Not Suitable for LUT-III :

Unit H11, H12, H21, H22, H31, H41, H42, H5, H6 & G31 are characterised by one or many qualities such as excessive relief, excessively drained, strongly sloping and very high & long duration flooding. Thus, these units are not suitable for cultivation in present situation, as these constraints cannot be modified even after incur of high cost for technology involved in alternation, modification and change in long term perspective, Such units are not suitable for this crop land use in near future.

LAND SUITABILITY FOR TURMERIC AND CASTER (LUT-IV) :

It is remarkable that mapping units with land qualities suitable for growth of turmeric and caster resemble similarity with LUT-III while otherwise is true in case of LUT-I and intermediate conditions are true for LUT-II. It may be mentioned that 43,999 ha land is suitable for this LUT which constitute 55.22% of the total area, though with its varying degree of suitability.

Highly Suitable Sub-Class :

These crops require specific conditions which are met in G11, G21 and G22, hence this LUT may well be praactised. These land units exhibit loam texture, well drained condition, high inherent fertility, 6.5-7.5 pH, normal relief, level land, high to medium organic carbon, very deep to deep soil, none to slight erosion and very low risk of flooding which are essential for its growth. Thus, these units are highly suitable for this LUT which spread over 26,547 ha and constitute 33.32% of the area.

Moderately Suitable Sub-Class :

Unit H32 and G12 are moderately suitable for this land utilization. Unit H32 is not that congenial due to unfavourable condition, namely clayey and light loam exture, medium CEC, 7.5-8.0 / 6.0-6.5 pH, gently sloping and medium to low organic matter whereas G12 owing to moderate erosion hazard, otherwise rests of the land qualities are highly suitable for this LUT. These units are spread over 1,142 ha which consists 1.43% of its total area. Chemical amendments and soil conservation measures may ameliorate its suitability to a considerable extent.

Marginally Suitable Sub-Class :

Land units, namely, G23, B11, A11, A12, A3, L1 and L2 are marginally suitable for this land utilization type. These units occupy 16,306 ha and constitute 20,47% of the total area (Table 18).

G23 is severely affected with erosion hazard while B11 is burdened with conditions marginally suitable for this LUT such as imperfectly drained, and clay loam to clay soil. Whereas, other land qualities of these mapping units are either highly or moderately suitable for this LUT.

The mapping unit A11 and A12 are burdened with severe limitations, namely sandy texture and imperfectly drained condition and moderate limitations in form of undulating slope and moderately deep soil. Rests of the land qualities are highly suitable for this LUT. As limitations mentioned above cannot be improved, these units remain under this class.

A3 unit is characterised with severe erosion hazard in addition to moderate limitation in form of clayey and light loam texture, medium inherent fertility, 7.5-8.0 and 6.0-6.5 pH, subnormal relief, gently sloping, medium to low organic matter and low risk of flood and comes under this suitability sub-class

Table 18
Rating Of Land Qualities Of Various Mapping Units for Turmeric and Caster (LUT - IV)

Sl. No.	Mapping unit	Texture	Drainage	Inherent fertility(CEC)	pH	Relief	Slope
1	2	3	4	5	6	7	8
1.	H11*	S2	N	S2	S1	N	S2
2.	H12*	S2	N	S2	S1	N	S2
3.	H21*	S2	N	S2	S1	N	N
4.	H22*	S1	N	S2	S1	N	N
5.	H31	S2	N	S2	S1	N	S2
6.	H32	S2	S1	S2	S2	S1	S2
7.	H41*	S2	N	S1	S2	S3	S1
8.	H42	S2	N	S1	S2	S3	S1
9.	H5	S2	N	S2	S1	N	N
10.	H6	S2	N	S1	S2	N	N
11.	G11	S1	S1	S1	S1	S1	S1
12.	G12	S1	S1	S1	S1	S1	S1
13.	G21	S1	S1	S1	S1	S1	S1
14.	G22	S1	S1	S1	S1	S1	S1
15.	G23	S2	S1	S2	S1	S2	S2
16.	G31*	S2	N	S1	S2	S3	S1
17.	G32	S2	N	S1	S2	S3	S1
18.	B11	S3	S3	S1	S2	S1	S1
19.	B12	S3	S3	S1	S2	S1	S1
20.	B13	S3	S2	S1	S2	S2	S1
21.	B2	S2	N	S1	S2	S3	S1
22.	A11*	S3	S3	S1	S2	S1	S2
23.	A12*	S3	S3	S1	S2	S1	S2
24.	A21	S3	N	S1	S2	S2	S1
25.	A22	S3	N	S1	S2	S2	S1
26.	A3	S2	S1	S2	S2	S2	S2
27.	L1	S3	S3	S1	S2	S1	S1
28.	L2	S3	S3	S1	S2	S1	S1

Contd. p-74

Organic matter	Soil depth	Erosion hazard	Risk of flooding	Suitability Classification		
				Order	Class	Sub- Class
9	10	11	12	13	14	15
S1	S1	S2	S1	N	N	NDr
S1	S1	S3	S1	N	N	NDr
S1	S2	S3	S1	N	N	NDrs
S1	S1	S3	S1	N	N	NDrs
S2	S1	S1	N	N	N	NDrsw
S2	S1	S1	S1	S	S2	S2fso
S2	S1	S1	N	N	N	NDw
S2	S1	S1	N	N	N	NDw
S2	S1	S3	S1	N	N	NDrs
S2	S1	S3	S1	N	N	NDrs
S1	S1	S1	S1	S	S1	S1
S1	S1	S2	S1	S	S2	S2e
S1	S1	S1	S1	S	S1	S1
S1	S1	S1	S1	S	S1	S1
S2	S1	S3	S1	S	S3	S3e
S1	S1	S1	N	N	N	NDw
S1	S1	S1	N	N	N	NDw
S1	S1	S1	S1	S	S3	S2Dt
S1	S1	S1	N	N	N	Nw
S1	S1	S1	N	N	N	Nw
S1	S1	S1	N	N	N	NDw
S1	S1	S1	S1	S	S3	S3D
S1	S1	S1	S1	S	S3	S3tD
S1	S1	S1	N	N	N	NDw
S1	S1	S1	N	N	N	NDw
S2	S1	S3	S2	S	S3	S3e
S2	S1	S1	S1	S	S3	S3Dt
S2	S1	S2	S1	S	S3	S3Dte

* These units are observed and mapped in the adjoining Bodhan area.

Mapping unit L1 and L2 are burdened with severe limitations, namely sandy texture and imperfectly drained condition to qualify for its marginal suitability for this LUT. In addition to these, there are moderate limitations such as 7.5-8.0 and 6.0-6.5 pH, medium to low organic matter and none to slight erosion in these units.

Not Suitable for LUT-IV :

H11, H12, H21, H22, H31, H41, H42, H5, H6, G31, G32, B12, B13, B2, A21 and A22 units exhibit at least one permanent constraint, namely, drainage, relief, slope or risk of flooding and make those units unsuitable for this land utilization. These units are spread over 27,211 ha and constitue 34.16% of the total area (Table 19).

Table 19

Land Suitability Categories And Summary Of Mapping Units Under Each Category For All LUTs

Suitability categories	Land Utilization Types (LUTs)			
	I	II	III	IV
Highly Suitable (S1)	A22, B13 & B2 (11,911 ha 14.95%)	B11 & B12 (2,845 ha 3.57%)	G21 & G22 (26,425 ha 33.17%)	G11, G21 & G22 (26,547 ha 33.32%)
Moderately Suitable (S2)	H42, G21, G22, G32 B11, B12, A11, A12, L1 & L2 (42,279 ha 53.07%)	G21, G22, B13, A11, A12, A21, A22. L1 & L2 (34,884 ha (44.53%)	H32, G11, G12, B11, B12, A11, A12, A21,A22 L1 & L2 (11,993 ha 15.04%)	H32 & G12 (1,142 ha 1.43%)
Marginally Suitable (S3)	H32, H41, G23, G31 & A3 (13,145 ha 16.50%)	H32, G23, G31, G32, B2 & A3 (27,537 ha 34.56%)	G23, G32, B13, B2 & A3 (29,190 ha 26.65%)	G23, B11, A11, A12, A3, L1 & L2 (16,309 ha 20.47%)
Not suitable (N)	H11, H12, H21, H22, H31, H5, H6, H11 & G12 (3,874 ha 4.86%)	H11, H12, H21, H22, H31, H41, H42, H5,H6 G11 & G12 (4,176 ha 5.24%)	H11, H12, H21, H22, H31, H41, H42, H5, H6 & G31 (3,601 ha 4.52%)	H11, H12, H21, H22, H31, H41, H42, H5, H6, G31, G32, B12, B13, B2, A21 & A22 (27,211 ha 34.16%)

Table 20
Relative Suitability, Potential LUTs And Irrigability Of Different Mapping Units

Mapping Units	Area ha	Area %	LUT I	LUT II	LUT III	LUT IV	Potential LUT	Land Irrigability
H11⊕	....	....	N	N	N	N	-	6st
H12⊕	---	---	N	N	N	N	-	6st
H21⊕	---	---	N	N	N	N	-	6st
H22	2,506	3.14	N	N	N	N	-	6st
H31	1,060	1.33	N	N	N	N	-	4st
H32	689	0.86	S3	S3	S2	S2	III,IV	3st
H41⊕	---	---	S3	N	N	N	-	5d
H42	302	0.38	S2	N	N	N	-	3d
H5	597	0.75	N	N	N	N	N	6st
H6	196	0.25	N	N	N	N	N	6st
G11	122	0.15	N	N	S2	S1	IV, III	2s
G12	453	0.57	N	N	S2	S2	IV, III	2s
G21	20,384	25.60	S2	S2	S1	S1	III ,IV	1
G22	6,041	7.58	S2	S2	S1	S1	III, IV	2s
G23	10,893	13.68	S3	S3	S3	S3	N	st
G31⊕	---	---	S3	S3	N	N	N	5std
G32	4,462	8.11	S2	S3	S3	N	I	2sd
B11	1,090	1.37	S2	S3	S3	N	I	2sd
B12	1,755	2.20	S1	S1	S2	S3	I,II	2sd
B13	2,342	2.94	S2	S1	S2	N	II, I	2sd
B2	7,930	9.96	S1	S2	S2	N	I, II	2sd
A11⊕	---	---	S1	S3	S3	N	II	1
A12⊕	---	---	S2	S2	S2	S3	II, III	2sd
A21	3,482	4.37	S2	S2	S2	S3	II, I	2sd
A22	1,639	2.06	S1	S2	S2	N	I, II	1
A3	1,563	1.96	S3	S3	S3	S3	-	3sd
L1	1,431	1.79	S2	S2	S2	S3	II, III	3sd
L2	1,332	1.67	S2	S2	S2	S3	II, III	3sd
Tanks	4,996	6.27	—	—	—	—	—	—
Misc.	2,403	3.01						

⊕ These mapping units are observed and mapped in adjoining Bodhan area.

RELATIVE SUITABILITY FOR POTENTIAL LUTs

Table vividly reveals that mapping unit H11, H12, H21, H22, H31, H41, H5, H6, G31 and A3 are not suitable for any of the land utilization considered for the survey area. These units may well be put under drought resistant vegetation to improve the ecosystem of area.

By virtue of its land qualities suitable for paddy cultivation and land irrigability, unit B12, B2, A11, A12 and A22 are highly suitable for paddy cultivation. Unit G23, B13, A11, A21, A22, L1 and L2 are suitable for LUT-II and III. Unit H32, G21 and G22 may be used for either LUT-II or IV. Unit G11 and G21 may be brought under LUT-IV or III taking into consideration of the land irrigability class. Unit B12 may be put either for LUT I or II. A12 may be put under LUT II or III (Table 19 and 20).

The author is pretty sure that if such type of land utilization is brought under consideration and practice depending upon its land irrigability based upon environmental conditions, a competitive and healthier crop production may take place and higher per ha yield may be obtained in days to come. In turn, it may ameliorate the social well being of people and area and contribute to increase in savings for investment in non-primary activities. At the same time, constant crop land planning, execution and management need to be taken care to achieve positive results while encouraging intense demonstration, subsidy, crop insurance, co-operative marketing etc, for fuller achievement of its objective in due course of time.

CHAPTER - 5

LAND IRRIGABILITY CLASSIFICATION

The interpretation of soil and land characteristics for irrigation, is primarily concerned with prediction of behaviour of soils under greatly altered water regime, brought about by the introduction of canal irrigation in the command area.

Irrigation projects, most often require special interpretation and classification of soils for its sustained use under irrigation. The land irrigability classification is based upon prevalent moisture holding capacity, effective rooting depth and intake characteristics of the soil. Soils are first grouped into soil irrigability classes according to their limitations for sustained use under irrigation, regardless of their location or the size of individual units.

SOIL IRRIGABILITY CLASSIFICATION

Soil irrigability classes are defined in terms of soil properties that reveal their degree of limitation for development and the requirement for irrigation management. The criteria for each class has quantitatively been defined whereas the most limiting properties of the soil determine the irrigability class.

Soil irrigability classes are very useful to group soils according to their suitability for its sustained use under irrigation. These classes are established irrespective of the availability of water for irrigation, water quality, land preparation costs, availability of drainage outfalls and other factors not related to soil.

Definitions of the five soil irrigability classes are mentioned below :

Class	Definition
Class A	None to slight soil limitations for sustained use under irrigation.

Class B	Moderate soil limitations for sustained use under irrigation.
Class C	Severe soil limitations for sustained use under irrigation.
Class D	Very severe soil limitations for sustained use under irrigation.
Class E	Not suited for irrigation

LAND IRRIGABILITY CLASSES

The suitability of land for irrigation mainly depends upon the physical characteristics of land unit, socio-economic factors of the area and cultural habits of the population in addition to soil irrigability classes. The more important factors for such a consideration are three, namely :

1. Quality and quantity of water.
2. Drainage requirements.
3. Other economic considerations.

Definitions of land irrigability classes are extracted below :

Class	Definition
Class 1	Land units that have few limitations of soil, topography or drainage for its sustained use under irrigation.
Class 2	Land units that have moderate limitations of soil, topography or drainage for its sustained use under irrigation.
Class 3	Land units that have severe limitations of soil, topography or drainage for its sustained use under irrigation.
Class 4	Land units that are marginal for its sustained use under irrigation because of very severe limitations of either soil, topography or drainage.

SOIL IRRIGABILITY CLASSES

In order to classify units into soil suitability classes which are based on soil properties, namely, effective soil depth (useful for crops), texture of surface soil of 30 cm., soil permeability, water holding capacity to a depth of 90 cm., coarse fragments percentage of pebbles and stones (above 75 mm.), gravel and *Kankar* (25 to 75 mm.), rock outcrops (distance apart in metres), salinity (in saturation extract), percentage of salt affected area, severity to alkali problem, sub-soil or substrata drainage characteristics and soil erosion status in arid and semi-arid conditions (Table 21)

Unit G11, G12, G21, G22 are characterised with Fine Loamy Udic Haplustalfs as they satisfy the criteria of class A soil irrigability. Besides these, A22 and A3 units also bear similar features. Soil depth goes more than 90 cm. and exhibit sandy loam to clay loam texture, 5.0-50 mm./hour permeability, more than 12 cm. water holding capacity, less than 5% coarse fragments, less than 15 mm. gravel and *kankar*, rock outcrops 40 metres apart, less than 4 mm. hos salinity, less than 20% visible salt affected area, less than 15% ESP, lower sub-soil is atleast moderately permeable and low level of soil erosion. Thus, these units are characterised with none to slight soil limitation for sustained use under irrigation. The wide range in land qualities of mapping units led them to qualify in class B as well.

Unit H11, H42, G11, G12, G21, G22, G31, G32 and B12 unlike earlier one, are less suitable for irrigation. These units are characterized with 45-90 cm. soil depth, loam surface texture, 1.5-5 mm./hour and 50-130 mm./hr soil permeability, 9-12 cm. water holding capacity, 5-15 coarse fragments, 15-35 mm. gravel and *kankar*, 20 metres apart rock outcrop, 4-8 mm. hos salinity, less than 20% area salt affected, ESP less than 15%, lower sub-soil moderately permeable and effect of rills led these units assume B class of soil Irrigability. These units have moderate limitations for its sustained use under irrigation.

Unit H31, H32, G23, B11, B12, B13, B2, A11, A12, A21, A22, L1 and L2 exhibit a wide range of soil characteristics at the family level. These are endowed with 22.5-45 cm. soil depth, sandy to clay surface texture, 0.3-1.3 mm. and 130-250 mm./hour soil permeability, 6-9 cm. water holding capacity, 15-35 mm. gravel and *kankar*, 15 metres apart rock outcrops, 8-12 mm. hos salinity, 20-50% soil affected area, ESP more than 15%, very low permeable sub-soil and, rills and gullies. Thus, these soils are burdened with severe limitations for its sustained use under irrigation and classed under C group.

Table 21
Criteria For Classifiying Soils Into Irrigability Classes In Semi-Arid And Arid Regions

S.No.	Soil Properties	Irrigable Soil Classes
		A
1	2	3
1.	Effective soil depth (useful to crops)	More than 90 cm.
2.	Texture of surface (30 cm.)	Sandy loam to clay loam inclusive
3.	Soil permeability*	5.0-50 mm/hr.
4.	Availble water holding capacity to depth of 90 cm.	12 cm. or more
5.	Coarse fragments (%) pebbles & stones (above 75 mm.)	Less than 5
6.	Gravel and *Kankar* (25 to 75 mm.)	Less than 15
7.	Rock out-crops (distance apart in m.)	40
8.	Salinity E.Cx10³ (in saturation extract)**	Less than 4 mmhos/
9.	Salt affected (visual) (% of area affected)	Less than 20
10.	Severity of alkali problem	ESP less than 15%
11.	Sub-soil or sub-strata drainage characteristics	Lower sub-soil is at least moderately permeableor a permeable layer of atleast 15 cm. thicknessoccurs immediately below the soil but within3 metre (sand, gravel)
12.	Soil erosion status	Effects of sheet and rill erosion are reflected in effective soil depth, available moisture holding capacity and some other factors are shown above. Moderate to severely gullied soils may be classified based on local experience.

Contd. p-82

Irrigable Soil Classes			
B	C	D	Non-Irrigable Soil Classes
4	5	6	7
45-90 cm. Loam	22.5-45 cm. Sandy, Clay	7.5-22.5 cm. Sandy, Clay	Less than 7.5 cm. Any texture
1.3-5mm/hr. & 50-130 mm/hr.	0.3-1.3 mm. & 150-250 mm/hr.	Less than 0.3 mm. & greater than 250 mm.	Not Applicable
9-12 cm.	6-9 cm.	2-6 cm.	Less than 2 cm.
5-15	15-35	35-65	More than 65
15-35	35-55	55-70	More than 70
20	15	5	Less than 5
4-8 mmhos	8-12 mmhos	12-16 mmhos	More than 16 mmhos
Less than 20	20-50	20-50	More than 50
ESP less than 15%	ESP more than 15%	ESP more than 15%	---------
No moderately permeable sub-soil or other permeable layer of atleast 15 cm. thickness occurs within depth of 3 metre.			

* Soil permeability as a criteria is not applicable to deep black soils because of their unique properties. Deep black soils (Vertisols) which are inherently slowly permeable due to expanding 2 : 1 lattice type minerals do not qualify for irrigability soil Class A. They would qualify for being placed in E, C and D class.

** The method recommended by the Soil Testing Laboratories in India prescribe 1 : 2 soil to water ratio for soil salinity determination and hence corresponding conductivity figures are given here.

Salinity in 1 : 2 dilution	Less than 1 mmhos	1-1.5mmhos
1.5-2.5 mmhos	2.5-3 mmhos	More than 3 mmhos

(The criteria for salinity and alkanity refer to equilibrium conductivity under irrigation with specified irrigation water)

H11 and H12 units are comprised with soils which are endowed with 7.5-22.5 cm. effective soil depth, sandy to clay surface texture, less than 0.3 mm. and greater than 250 mm./hour permeability, 2-6 cm. water holding capacity, 35-65 mm. coarse fragments, 55-70 mm. gravel and *kankar*, 5 metres apart rock out-crops, 12-16 mm. hos salinity, 20-50% salt affected area, ESP more than 15%, very low/no sub-soil permeability and moderate to severely gullied and eroded soil characteristics. Thus, these are classes as D from the view point of soil irrigability.

Unit H21, H22, H5 and H6 are characterized with soil depth less than 7.5 cm., any surface texture not suitable for irrigation, less than 2 cm. water holding capacity, more than 65 mm. coarse fragments, more than 70 mm. gravel and *kankar*, less than 5 metres apart rock out-crop, more than 16 mm. hos salinity, more than 50% salt affected area, non-permeable sub-surface layer below 15 cm. and severely subject to gullies and erosion. Thus, these soils are not suitable for irrigation in the area and put under E class.

LAND IRRIGABILITY CLASSIFICATION

Land irrigability classification has been done on the basis of its specification required. In this regard, soil irrigability class; slope and surface grading of topography; and outlets, - surface, sub-surface and depth of water table components of drainage have been taken into consideration. Accordingly lands of the study area have been classified under following land irrigability classes :

Class 1 : Lands under this class have no limitations for sustained use under irrigation. Land units, which have 'A' soil irrigability class, less than 1% slope, free from grading restrictions, suitable outlets available, free from sub-surface drainage and depth of water table more than 5 metre, are ideal areas for irrigation. G21, B2 and A22 mapping units come under the category which cover an area of 29,453 ha and constitute 37.58% of the total area (Table 22).

Class 2 : Lands under this class have moderate limitations for sustained use under irrigation. Major limitations of these lands soil irrigability, moderate permeability and drainage. This class includes B11, B12, B13, A11, A12 and A21 which occupy 21,747 ha and represent 27.29% of the total area.

Table 22
Specifications For Land Irribability Classes

Land Characteristics	Irrigable Land Class Class 1	Class 2	Class 3	Class 4	Class 5 Temporarily non-irrigable (unclassified)	Class 6 Not suitable for irrigation
1	2	3	4	5	6	7
			SOILS			
Soil Irrigability Class	A	A to B	A to C	A to D	Further investigations needed	Includes lands which do not meet the minimum requirement for other land classes and are not suitable for irrigation or small isolated tracts (specifying size or distance from canal) not susceptible to delivery of irrigation water.
			TOPOGRAPHY			
1. Slope	Less than 1%	1-3%	3-5%	5-10%	-----	
2. Surface grading	No restrictions	Moderate restrictions	Moderate to severe restrictions	Severe restrictions	-----	
			DRAINAGE			
1. Outlets	Suitable outlets available	Suitable outlets available	Suitable outlets available	No drainage outlets available	Further investigations needed	
2. Surface	(Specifications to be developed for each project area).					
3. Subsurface	No subsurface drainage needed	No subsurface drainage needed	Subsurface drainage needed	No natural drainage outlets		
4. Depth of water table	More than 5 Meter	3-5 metre	1.5-3 metre	1.5 metre and below		

With regard to items under Topography (1) & Drainage (2) & (3), the criteria will have to be worked out for each project on the basis of local conditions.

Class 3 : Land units that have severe limitations for its sustained use under irrigation are classed under this category. The topography includes complex slope and undulating terrain. Other limitations areerosion, poor permeability and poor drainage. H32, H41, G23, A3, L1 and L2 units occupy an area of 16,210 ha which represent 20.32% of the area.

Class 4 : The land units which are marginal for its sustained use under irrigation, are classed under this category. The severe limitation of this class of irrigability include soil(s) and topography (t). H31 is the land unit which falls under this category. It occupies an area of 1,060 ha and constitutes 1.33% of the total area (Table 23).

Class 5 : The land units temporarily not suitable for its sustained use under irrigation are included under this category. There are very severe limitation of topography and moisture. H41 and G31 are the units which have been observed in the adjoining area but not reported in this part of study area (See Table 24).

Class 6 : The land units not suitable for its sustained use under irrigation are grouped under this category. These are characterized with very severe limitations of soil, topography and excessive drainage. H11, H12, H21, H22, H5 and H6 mapping units happen to occur under this class and confine in the hill landforms. These units are spread over an area of 4,359 ha and represent 5.46% of the total area.

Table 23
Physiography, Soils and Irrigability Classification

S.No.	Mapping Units	Physiography
1	2	3
1.	H11*	Hill crest, moderately dense forests
2.	H12*	Hill crest, sparse forests
3.	H21*	Hill side slope, moderatelv dense forests
4.	H22	Hill side slope, sparse forest
5.	H31	Foot slope, sparse vegetation
6.	H32	Foot slopes, cultivated
7.	H41*	Hill Valley, submerged
8.	H42	Hill valley, cultivated
9.	H5	Inselberg, sparse vegetation
10.	H6	Dyke, sparse vegetation
11.	G11	Crest, Granitic Pediment plain cultivated
12.	G12	Crest, Pediment plain waste land
13.	G21	Side slope, pediment plain, terraced, cultivated
14.	G22	Side slope, pediment plain, unterraced, cultivated
15.	G23	Side slope of pediment plain, waste land
16.	G31*	Pediment valley, cultivated
17.	G32	Pediment valley cultivated
18.	B11	Buried Pediment, large parcelling
19.	B12	Buried Pediment, medium parcelling
20.	B13	Buried pediment, small parcelling
21.	B 2	Broad valley, paddy cultivation
22.	A11*	Gently sloping, alluvial plain, large parcelling
23.	A12*	Gently sloping, alluvial plain, small parcelling
24.	A21	Nearly level, alluvial plain, large parcelling
25.	A22	Nearly level, alluvial plain, small parcelling
26.	A3	Sloping lands of interfluve, waste land
27.	L1	Limestone landscape, Cultivated
28.	L2	Limestone landscape, Uncultivated

Soil Association	Soil Irrigability Class	Land Irrigability Class	Remarks
4	5	6	7
1. Fine Loamy Udic Ustochrepts	D	6st	Due to high elevation of landform
1. Loamy Skeletal Udic Haplustalfs	D	6st	-do-
2. Loamy Skeletal Typic Paleustalfs	D	6st	-do-
1. Fine Loamy Typic Haplustalfs	E	6st	Due to high elevation steep slopes & severe rocky & stony phase
2. Fine Loamy Udic Haplustalfs	E	6st	-do-
1. Loamy Skeletal Lithic Ustorchents	E	6st	-do-
2. Fine Loamy Udic Haplustalfs	E	6st	-do-
1. Loamy Skeletal Udic Haplustalfs	C	4st	
2. Fine Loamy Udic Haplustalfs	C	4st	
1. Fine Loamy Udic Haplustalfs	C	3st	
2. Fine Loamy Udic Ustochrepts	C	3st	
1. Fine Loamy Typic Ustochrepts	B	3d	Concave landform
1. Fine Loamy Typic Ustochrepts	B	3d	Concave landform
1. Loamy Skeletal Udic Haplustalfs	E	6st	Due to high elevation of landform
1. Fragmental Udic Haplustalfs	E	6st	Due to high elevation of landform
1. Loamy Skeletal Udic Haplustalfs	B	2s	
2. Fine Loamy Udic Haplustalfs	A	1	
1. Loamy Skeletal Udic Haplustalfs	B	2s	
2. Fine Loamy Udic Haplustalfs	A	1	
1. Fine Loamy Udic Haplustalfs	A	1	
2. Loamy Skeletal Udic Haplustalfs	B	2s	
1. Loamy Skeletal Udic Haplustalfs	B	2s	
2. Fine Loamy Udic Haplustalfs	A	1	
1. Loamy Skeletal Udic Haplustalfs	C	3st	
1. Fine Aquic Ustochrepts	B	5std	Concave landform used for water storage
1. Fine loamy Aeric Haplaquepts	B	2sd	
2. Fine Loamy Fluventic Ustochrepts	B	2sd	
1. Fine Typic Chromusterts	C	2sd	
2. Fine Vertic Ustochrepts	C	2sd	
1. Fine Vertic Ustochrepts	C	2sd	
2. Fine Loamy Typic Ustochrepts	B	2s	
1. Fine Vertic Ustochrepts	C	2sd	
1. Fine Vertic Haplaquepts	C	2sd	
2. Fine Loamy Typic Haplaquepts	B	2s	
1. Fine Typic Chromusterts	C	2sd	
1. Fine Typic Chromusterts	C	2sd	
1. Fine Typic Chromusterts	C	2sd	
1. Fine Vertic Ustochrepts	A	1	
2. Fine Typic Chromusterts	C	2sd	
1. Fine Vertic Ustochrepts	C	3st	
1. Fine Typic Chromusterts	C	3sd	
2. Fine Typic Pellusterts	C	3sd	
1. Fine Typic Chromusterts	C	3sd	

* The H11, H12, H21, H41, G31, A11, A12 mapping units are found and mapped only in adjacent Bodhan area

Table 24
Area Under Different Irrigability Classes

Irrigability Class	Category Subclass	Mapping Units
1	---------	G21, B2, A22
2	2s	G11, G12
	2sd	G22, G32, B11, B12, B13, A11*, A12, A21
3	3d	H42
	3sd	L1, L2
	3st	H32, G23, A3
4.	4st	H31*
5	5d	H41*
	5std	G31
6	6st	H11*, H12*, H21*, H22, H5, H6

Contd. p-89

Area	
hr.	(%)
29,953	37.50
575	0.72
21,172	26.57
302	0.37
2,763	3.46
13,145	16.49
1,060	1.33
--	--
--	--
4,359	5.46

* These mapping units are found and mapped only in adjoining Bodhan area.

CHAPTER - 6

SUMMARY AND CONCLUSION

The survey area is a part of Nizamsagar Command which occupies north-western part of Telangana Pediplain. The acute siltation process reduced the water storage capacity of Nizamsagar thus, aims and objectives of the command project remained unfulfilled. Keeping in view the modernization of the command area by adopting best possible land use and agronomic practices according to the land suitability, the present work was taken up. Semi-detailed soil survey was conducted using panchromatic aerial photographs on a scale of approximately 1 : 20,000 with selective field checks.

Pre-field interpretation was done and tentative legend was prepared which was modified after reconnaissance survey wherever it became necessary. Three sample strips were selected and studied intensively and selective field checks were made later on. The area was divided into 5 main land types, namely, Hills, Pediment plain (Granitic), Buried Pediment and Limestone, landscape which were further sub-divided on the basis of their different physiographic attributes and land uses. Different phases of soils were also worked out for dominant soils in each physiographic sub-unit. A physiography-cum-soil map was prepared on 1 : 20,000 and interpreted for land irrigability and soil suitability for various land uses.

Considering the agro-climatic conditions, topography, physical and chemical properties of the soils, and related soil problems, the mapping units of survey area have been classified into suitability classes, namely, Highly suitable, Moderately suitable, Marginally suitable and Not suitable for four Land Utilization Types, namely, LUT-I Paddy cultivation, LUT-II Sugarcane, LUT-III Jowar, maize and millets, and LUT-IV Turmeric and Caster crops.

A22, B13 and B2 units (11,911 ha and 14.95%) are highly suitable, H42, G21, G22, G23, B11, B12, A11, A12, A21, L1 and L2 are moderately suitable (42,279 ha and 53.07%), H32, H41, G23, G31 and A3 are marginally suitable (13,145 ha. and 16.50%) and H11, H12, H21,

H22, H31, H5, H6, G11 and G12 are not suitable (3,874 ha. 4.86%) for Land Utilization Type - I, i.e. Paddy cultivation.

B11 and B12 units (2,845 ha. and 3.57%) are highly suitable, B13, G21, G22, A11, A12, A21, A22, L1 and L2 are moderately suitable (34,884 ha. and 44.53%), H32, G23, G31, G32, B2 and A3 units are marginally suitable and H11, H12, H21, H22, H31, H41, H42, H5, H6, G11 and G12 are not suitable (4,176 ha. and 5.24%) Land Utiliztion Type-II, i.e. Sugarcane cultivation.

G21 and G22 mapping units (26,425 ha. and 33.17%) are higly suitable H32, G11, G12, B11, B12, A11, A12, A21, A22, L1 and L2 units (11,993 ha. and 15.04%) are moderately suitable, G23, G32, B13, B2 and A3 units (29,190 ha. and 36.65%) are marginally suitable and H11, H12, H21, H22, H31, H41, H42, H5, H6 and G31 units (3,601 ha. and 4.52%) are not suitable for Land Utilization Type-III, i.e. Jowar, Maize, Millets.

G11, G21, and G22 mapping units (26,547 ha. and 33.32%) of the area are highly suitable, H32 and G12 (1,142 ha. and 1.43%), moderately suitable, G23, B11, A11, A12, A3, L1 and L2 (16,309 ha. and 20.47%) marginally suitable and H11, H12, H21, H22, H31, H41, H42, H5, H6, G31, G32, B12, B13, B2, A21 and A22 (27,211 ha. and 34.16%) not suitable for Land Utilization Type-IV, namely, Turmeric and Caster.

Interpretation of soil map for land irrigability has also been done under six categories. Class 1 land has no limitation for sustained use under irrigation thus ideal for irrigation. Under this irrigability class G21, B2 and A22 units occur which comprise of 29.953 ha. and 37.58% of the total. Class 2 lands have moderate limitations for sustained use under irrigation with limitations of permeability and drainage. G11, G12, G22, G32, B11, B12, B13, A11, A12 and A21 come under this class. These comprise of 21,747 ha. and 27.29% of the total area. The Class 3 land has severe limitations for sustained use under irrigation. Complex slope, undulating terrain and erosion are major limitations of the said class. These units cover 16,210 ha. of area and constitute 20.32% of the surveyed area. H32, H42, G23, A3, L1 and L2 mapping units come under this class.

Class 4 lands have very severe limitations of soil, topography and erosion. H31 unit is classed in this class which occupies an area 1,060 ha. and consists of 1.33% of the total area.

Class 5 lands are temporarily not suitable for sustained irrigation. H41 and G31 come under this category. Class 6 lands are not suitable for sustained use under water. There are severe limitations of soil, topography, excessive drainage, etc. H11, H12, H21, H22, H5 and H6 mapping units which cover 4,359 ha. and represent 5.46% of the total area.

Land Utilization according to the suitability of soil and irrigation according to land irrigability may fulfil the needs and requirements of rural population of the area by increasing yield per unit area along with least disturbances in the ecological balance.

The advances in the field of remote sensing have greatly enhanced the capability of geoscientists in making rapid, more reliable and up-to-date thematic maps of natural resources. Recent advancements and sophistication in related field of science and technology have contributed tremendous refinements and developments in Multispectral sensors, better resolution of images, quick repeatitive coverage, automated cartography, land information system etc. essential for rapid and accurate comprehension of natural resources and environment. Thus, remote scientists have to make use of their tools and techniques in qualitative improvement in social well-being of man kind on the one side and to contribute restoration of eco-order in the ecosystem and environment of the mother Earth.

BIBLIOGRAPHY

1. **Bennema, J. and Gelens, N.F. (1969) :**
   Aerial Photo-interpretaion for Soil Surveys, I.T.C. Lecture Notes, I.T.C. Delft.

2. **Buringh, P. (1960) :**
   The application of Aerial Photography in Soil Surveys, Manual of Photographic Interpretation, Washington, U.S.A.

3. **Donald, Steila (1976) :**
   The Geography of Soils, Prentice- Hall, Inc. Englewood Cliff, New Jersey, 135-146.

4. **FAO (1961) :**
   Guidelines of Soil Profile Description, Rome, Italy.

5. **FAO (1976) :**
   A Framework for Land Evaluation, F.A.O., Rome, Italy.

6. **Goosen, D. (1967) :**
   Aerial Photo-interpretation in Soil Surveys, Soil Bull. 6, FAO, Rome, Italy.

7. **Hilwig, F.W. (1980) :**
   Visual Interpretation of Multi-Temporal Landsat Data for Inventories of Natural Resources, I.T.C. Journals, 297-327.

8. **Janny, H. (1941) :**
   Factors of Soil Formation, McGraw Hill, New York.

9. **John, R.H. (1977) :**
   Applied Geomorphology, Elsevier Scientific Publication Company, Amsterdam, 277-286.

10. **Manchanda, M.L. (1981) :**
Study of Landsat Imageries and Aerial Photos for Evaluation of Salt-Affected Soils of part of North West India, I.T.C. Enschede, The Netherlands.

11. **Manual of Remote Sensing (1983) :**
American Society of Photogrammetry (ed : Robert N. Colwell), 2111-2162.

12. **Piper, C.S. (1957) :**
Soils and Plant Analysis, Hans Publications, Bombay.

13. **Samacharya, K. and T.R. Srinivasan (1972) :**
Use of Small Scale Aerial Photographs for Preparation of Small Scale Soil Map, paper presented at 12th Congress of International Society of Photogrammetry.

14. **Simonson, R.W. (1959) :**
Outline of Generalized Theory of Soil Genesis, Soil Science Society, America Proceedings, 23, 152-56.

15. **Singh, R.L. (ed. 1971) :**
India : A Regional Geography, NGSI, Varanasi, 1971.

16. **Soil Survey Manual (1970) :**
A.I.S. & L.U.S., New Delhi.

17. **Soil Survey Staff (1962) :**
Soil Survey Manual, Handbook No. 18, U.S.D.A., Washington.

18. **U.S.D.A. (1975) :**
Soil Taxonomy : A Basic System of Soil Classification for Making and Interpreting Soil Surveys, U.S.D.A. Handbook No. 436.

19. **Vink, A.P.A. (1963) :**
Aerial Photographs and Soil Sciences, NNESCO Proc. Toulouse Cont.

20. **Vink, A.P.A. (1963) :**
Planning Soil Survey in Land Department, I. T. C. Pub. 10, Wageningen.

Pedon No. 1

FRAGMENTAL MIXED HYPERTHERMIC UDIC HAPLUSTALFS

These are very deep, excessively to well drained, dark reddish brown soils developed on dyke near Armur. Soils are *in situ* formed and are derived from basic rocks.

Location	:	About 500 m. north from Armur on linear feature crest (dyke).
Physiographic position	:	Linear feature (Dyke).
Drainage class	:	Excessively drained.
Land use	:	Sparse Vegetation
Parent material	:	Granitic gneiss.
Sampled by	:	D.R. Singh and R.S. Yadava
Profile No.	:	62
Date of examination	:	22.2.1985

Horizon	Depth (cm.)	Description
B21t	0-19	Dark reddish brown (5 YR 3/2); silty clay; medium subangular blocky structure; hard (dry) firm (moist) plastic (wet); few medium and coarse and many very fine and fine pores; patchy thin cutans on ped faces; few medium and many very fine and fine roots, no lime reaction; clear smooth boundary.
B22t	19-54	Dark reddish brown (5 YR 3/4); clayey medium subangular blocky structure; hard (Dry), firm (moist) sticky and plastic (wet); few medium and many very fine and fine pores; common patchy thin cutans on ped faces and pores; few medium and common very fine and fine roots; no lime reaction, clear smooth boundary.

B23t	54-75	Dark reddish brown (5YR 3/2); clayey; medium subangular blocky structure; hard (dry), firm(moist), sticky and plastic (wet); few medium and many very fine and fine pores; common patchy thin cutans on ped faces and pores; few medium and common very fine and fine roots; no lime reaction, clear smooth boundary.
B24t	75-90+	Dark reddish brown (5 YR 3/3.5); clayey; medium subangular blocky structure; firm (moist); very (sticky) and very plastic (wet); many very fine and fine pores; patchy thin cutans on ped faces and pores; few very fine roots; no lime reaction.

RANGE IN CHARACTERISTICS

The surface horizon is 15-25 cm. thick. No lime reaction was observed throughout the profile. Soils are very deep having rock fragments with soils.

LAND USE AND MANAGEMENT

Sparse vegetation is found in this unit. Pasture practices are important which led to degradation of forests. Soils are subject to moderate and severe erosion which require planned forestry practices for maintaining soils suitable for forest growth.

Pedon No. 2

FINE LOAMY MIXED HYPERTHERMIC UDIC HAPLUSTALFS

These soils are very deep, reddish brown, well drained, non-calcareous in nature. Soils are derived from granitic gneiss.

Location	:	1 km. from Armur
Physiographic position	:	Top portion of Pediment Plain.
Drainage class	:	Well drained.
Land use	:	Waste land subject to grazing.
Parent material	:	Granitic gneisses.
Sampled by	:	D.R. Singh and R.S. Yadava
Profile No.	:	82
Date of examination	:	27.2.1985

Horizon	Depth (cm.)	Description
A1	0-15	Dark reddish brown (5YR 3/3), moist; reddish brown (5 YR 5/3) dry; sandy loam; fine subangular blocky structure; slightly hard (dry); friable (moist) and slightly sticky and slightly plastic (wet); few medium common very fine and fine pores; common very fine and fine roots; no lime reaction; clear smooth boundary.
B1	15-34	Dark reddish brown (5YR 3/3.5) moist; clay loam; medium subangular blocky structure; friable (moist); sticky and slightly plastic (wet); few medium and many very fine and fine pores; few very fine roots; no lime reaction; clear smooth boundary.
B21t	34-58	Dark reddish brown (5 YR 3/3) moist; clay loam plus; medium subangular blocky structure; friable (moist), sticky

		and slightly plastic (wet); few coarse and many very fine and fine pores; patchy thin cutans on ped faces and pores; few very fine roots; no lime reaction; clear smooth boundary.
B22t	58-70	Dark reddish brown (2.5 YR 3/3) moist; clay loam plus; medium subangular blocky structure; friable (moist) and sticky and plastic (wet); many very fine and fine pores; patchy thin cutans on ped faces and pores; few medium and few very fine roots; no lime reaction; clear smooth boundary.
C	70-110+	Dark reddish brown (2.5 YR 3/4) moist; clay loam plus; medium subangular blocky structure; friable (moist), sticky and plastic (wet), many very fine and fine pores; no lime reaction; clear smooth boundary.

RANGE IN CHARACTERISTICS

The surface horizon is 15-20 cm. thick. No lime reaction was observed throughout the profile. Soils are very deep having skeletal material below 70 cm.

LAND USE AND MANAGEMENT

Grazing practices are done in this unit thus degradation of vegetation cover has taken place. These practices need to be discontinued and scientific forestry measures should be adopted in favour of maintaining the ecosystem of the area.

SOIL ASSOCIATION

Loamy skeletal Mixed Hyperthermic Udic Haplustalfs.

Pedon No. 3

FINE LOAMY MIXED HYPERTHERMIC AERIC HAPLAQUEPTS

These are very deep soils characterized with very dark grey to very dark greyish brown, moderately well drained, weak to strong lime reaction and developed in pediment valleys. Soils are derived from alluvium.

Location	:	About 125 m. north east from the Culvert in Pediment Valley, east of Anksapuram.
Physiographic position	:	Pediment valleys surrounded by pediment plain.
Drainage	:	Moderately well drained.
Land use	:	Cultivation
Parent material	:	Alluvium
Sampled by	:	D.R. Singh and R.S. Yadava
Profile No.	:	6
Date of examination	:	11.2.1985

Horizon	Depth (cm.)	Description
Ap1	0-11	Very dark greyish brown (10 YR 3/2), silty clay mottled with very pale brown (7.5 YR 4/4) material; moderate medium subangular blocky structure; hard (dry), friable (moist), plastic (wet); common very fine and fine pores; many very fine and fine roots; strong lime reaction; clear smooth boundary.
Ap2	11-25	Very dark greyish brown (10 YR 3/2); silty clay, mottled with very pale brown (7.5 YR 4/4) material, moderate medium subangular blocky structure ; firm (moist), very plastic (wet); few gravels; common very fine and fine pores; many very fine roots; stony lime reaction; clear and smooth boundary.

B21	25-46	Very dark greyish brown (10 YR 3/1.5) silty clay mottled with dark reddish brown (5 YR 3/4) material; medium subangular blocky to fine angular blocky structure, friable (moist) plastic (wet); many very fine and fine pores; common very fine roots; weak lime reaction; clear smooth boundary.
B22	46-56	Very dark greyish brown (10 YR 3/2); sandy clay loam mottled with reddish brown (5 YR 4/4) material; medium subangular blocky structure; friable (moist), plastic (wet); many very fine and fine pores; few very fine roots; weak lime reaction, clear smooth boundary.
B31	56-69	Dark brown (10YR 3/2.5), sandy loam; medium subangular blocky structure; friable (moist) non - plastic (wet), many very fine pores; few very fine roots; no lime reaction, clear and smooth boundary.
B32	69-89	Dark yellowish brown (10YR4/3.5); sandy loam; medium subangular blocky structure; friable (moist), non-plastic (wet); common very fine pores; few very fine roots; no lime reaction, clear smooth boundary.
C1	89-105	Greyish brown (2.5Y 4.5/2); sandy loam; medium subangular blocky structure; friable (moist), non-plastic (wet); few very fine pores; few very fine roots; weak lime reaction, clear smooth boundary.
C2	105-125	Greyish brown (2.5Y 4.5/2); sandy clay loam, massive structure; weak lime reaction (Auger).
C3	125-150	Dark greyish brown (2.5Y 4/2); sandy loam, strong lime reaction (Auger).

RANGE IN CHARACTERISTICS

The surface layer is generally 15-25 cm. thick, effervescence is weak to strong. Chroma ranges from 1.5 to 3.5 and value from 3 to 4 in hue of 10YR and 2.5Y.

LAND USE AND MANAGEMENT

These soils are mostly under paddy cultivation and are bunded for stagnating water for paddy cultivation.

ASSOCIATED SOILS

The soils are associated with Fine Loamy Mixed Hyperthermic Fluventic Ustochrepts.

Pedon No. 4

FINE MIXED HYPERTHERMIC VERTIC USTOCHREPTS

These soils are very deep, very dark greyish brown, moderately well drained, weakly to strongly calcareous, developed over valley floor of buried pediment.

Location	:	750 m. east of PWD Irrigation Bungalow
Physiographic position	:	Buried pediment valley
Drainage	:	Moderately well drained.
Land use	:	Paddy Cultivation
Parent material	:	Alluvium
Sampled by	:	D.R. Singh and R.S. Yadava
Profile No.	:	86
Date of examination	:	28.2.1985

Horizon	**Depth (cm.)**	**Description**
Ap	0-18	Very dark greyish brown (10 YR 3/2); moist; dark greyish brown (10YR 3.5/2) dry, silty clay loam; coarse subangular blocky; hard (dry), firm (moist), sticky and plastic (wet); few coarse, common medium and many very fine and fine pores; many very fine and fine roots; strong lime reactions; clear smooth boundary.
B21t	18-43	Dark brown (10YR3/2.5) moist, clay loam plus; coarse subangular blocky; firm (moist), sticky and plastic (wet); few medium and many very fine and fine roots; moderate lime reaction; clear smooth boundary.
B22t	43-61	Dark brown (10YR 3/3) moist; clay loam plus; medium subangular blocky struc-

		ture; firm (moist); sticky and very plastic (wet); few medium and many very fine and fine pores; common very fine roots; weak lime reaction; clear smooth boundary.
B23t	61-88	Dark brown (10YR 3/2.5) moist; clay loam plus; medium subangular blocky structure, firm (moist), sticky and plastic (wet); many very fine and fine pores; few very fine roots; weak lime reaction; clear smooth boundary.
B3	88-120+	Dark brown (10YR 3/3) moist; clay loam; medium subangular blocky structure; firm (moist), sticky and plastic (wet); many very fine and fine pores; few very fine roots; weak lime reaction.

RANGE IN CHARACTERISTICS

Surface horizon is 15-25 cm. thick, moderately well drained to somewhat imperfectly drained soils. Weak to strong lime reaction has been noticed throughout the profile. Soils are very deep and fertile in the area.

LAND USE AND MANAGEMENT

This type of soil is used for intensive paddy cultivation. Bunding and levelling has been done to take good paddy crops. Areas close to tanks are liable to waterlogging conditions.

ASSOCIATED SOILS

Fine Loamy Mixed Hyperthermic Typic Haplaquepts.

Pedon No. 5

FINE LOAMY MIXED HYPERTHERMIC TYPIC USTOCHREPTS

The soils belonging to this family are very deep, very dark brown to dark yellowish brown, moderate to strongly calcareous and well drained. They are found to occur on the valleys of granitic pediment plain.

Location	:	About 3 km. south of Warni village, 100 metres inside on the western side of the road.
Physiographic position	:	Valley of granitic pediment plain.
Drainage	:	Moderately well to well drained.
Land use	:	Cultivation
Vegetation	:	Teak, palas, thorny plants, creepers.
Parent material	:	Alluvium
Sampled by	:	S.C. Dhar, V.K.B. Sawant, S. Gopi
Profile No.	:	44
Date of examination	:	19.2.1985

Horizon	**Depth (cm.)**	**Description**
Ap	0-14	Very dark brown (10YR 2.5/2) moist; sandy clay loam; weak medium subangular blocky structure; soft (dry), friable (moist), sticky and plastic (wet); common fine and medium pores; common very fine and fine roots; clear smooth boundary.
B1	14-43	Dark yellowish brown (10YR 4/4) moist; heavy sandy clay loam; weak fine and medium subangular blocky structure; slightly hard (dry), friable (moist); very sticky & plastic (wet); common medium pores; common medium and coarse roots; weak effervescence ; gradual smooth boundary.

B21	43-60	Dark yellowish brown (10YR 4/6) moist; heavy sandy clay loam; about 15% coarse fragments of granitic composition; weak fine subangular blocky structure; friable (moist), very sticky and plastic (wet); very few fine and coarse roots; moderate effervescence; clear smooth boundary.
B22	61-105	Dark yellowish brown (10YR 4/6) moist; sandy clay loam; about 20% coarse fragments of granitic composition; weak fine subangular block structure; friable(moist) very sticky and plastic (wet); few medium and coarse pores; common medium and coarse roots; strong effervescence; gradual smooth boundary.
C	105-170	Dark yellowish brown (10YR 4/4) moist; heavy sandy loam; about 30% coarse fragments of granitic composition; structureless; loose (moist), slightly sticky and slightly plastic (wet); few fine pores; violent effervescence.

RANGE IN CHARACTERISTICS

The surface horizon is generally 14-20 cm. thick. The colours vary from 10YR 2/2 to 10YR 6/4. The sub-surface texture varies from clay loam to silly clay loam and sandy clay. The distribution of lime throughout the profile depth varies from weak to strong and at places it is is violent.

LAND USE AND MANAGEMENT

These soils are generally cultivated for paddy. The cropping intensity is 100%. The fields are well managed.

ASSOCIATED SOILS

Fine Aquic Ustochrepts are associated with these soils.

Pedon No. 6

FINE MIXED HYPERTHERMIC TYPIC USTOCHREPTS

These soils are very deep, dark brown to dark greyish brown, calcareous and development over valleys of granitic pediment plain. These are moderately well to well drained and derived from alluvium.

Location	:	1/2 km south of Amdapur village
Physiographic position	:	Intensively cultivated valleys of granitic pediment plain.
Drainage	:	Moderately well to well drained
Parent material	:	Alluvium
Sampled by	:	S. Gopi, V.K.B. Sawant, S.C. Dhar
Profile No.	:	75
Date of exam.	:	27.2.1985

Horizon	**Depth (cm)**	**Description**
Ap	0-30	Dark brown (10 YR 4/3) moist; brown (10 YR 5/3) dry; silty clay; moderate fine subangular blocky structure; firm (moist), very sticky and very plastic (wet); common fine pores; few fine roots; weak effervescence; clear smooth boundary.
B1	30-50	Very dark greyish brown (10 YR 3/2) moist; sandy clay; moderate fine subangular blocky structure; firm (moist), very sticky and very plastic (wet); common fine pores; few fine roots; weak effervescence; clear smooth boundary.
B2	50-90	Dark brown (10 YR 3/3) moist; sandy clay; weak to moderate medium subangular blocky structure; firm (moist) very sticky and very plastic (wet); few fine pores; few fine roots; weak effervescence; clear smooth boundary.

B22	90-165	Dark brown (10 YR 4/3) moist; clay; weak to moderate subangular blocky structure; firm (moist); very sticky and very plastic (wet); few fine pores; weak effervescence.

RANGE IN CHARACTERISTICS

The surface colours vary from 10YR 4/3 to 2.5Y 4/2 and the textures are generally heavy clay loam and silty clay. The subsurface horizon is more than 100 cm deep varying in colours from 10YR 3/1.5 to 10Y 4/2. The subsurface textures are sandy clay and clay.

LAND USE AND MANAGEMENT

These soils are generally intensively cultivated for paddy where cropping intensity is 150%. The fields have generally small parceling and are properly levelled and bunded.

Pedon No. 7

FINE LOAMY MIXED HYPERTHERMIC AQUIC USTOCHREPTS

These soils are very deep, very dark greyish brown to dark greyish brown, moderately well drained, weakly to strongly calcareous, developed over buried pediment plain and foot-hill slopes.

Location	:	About 1/2 km. south east of Narullabad village, 200 meters away on the north side of the road leading to Bamandevpalli.
Physiographic position	:	Buried pediment plain
Drainage	:	Moderately well drained.
Land use	:	Cultivation and waste land with scanty vegetation.
Sampled by	:	S.C. Dhar, V.K.B. Sawant, S. Gopi
Profile No.	:	58
Date of examination	:	23.2.1985

Horizon	Depth (cm.)	Description
Ap	0-20	Very dark greyish brown (10YR 3.5/2) moist; dark greyish brown (2.5Y 5.5/2) dry; sandy clay loam; weak coarse subangular blocky structure; very hard (dry), firm (moist), sticky and plastic (wet); common fine and few medium pores; few fine roots; weak effervescence; clear smooth boundary.
B1	20-45	Very dark greyish brown (10YR 3/2) moist; sandy clay loam; massive structure; firm (moist); sticky and plastic (wet); common fine and few medium pores; few fine and medium roots; moderate effervescence; clear smooth boundary.
B21	45-90	Dark greyish brown (2.5Y 3.5/2) moist;

		heavy sandy clay loam; massive structure, firm (moist), very sticky and very plastic (wet); common medium distinct greyish brown (10YR 5/2) motles; few fine pores; moderate effervescence; gradual smooth boundary.
B22	90-120	Dark greyish brown (2.5Y 4/2) moist; heavy sandy clay loam; massive structure; firm (moist); very sticky and very plastic (wet); common medium distinct greyish brown (10YR 5/2) mottles; few fine pores; strong effervescence; clear smooth boundary.
B3	120-160	Dark greyish brown (2.5Y 4/2) and yellowish brown (10YR 5/4) moist; sandy clay loam; firm (moist); very sticky and very plastic (wet); strong effervescence.

RANGE IN CHARACTERISTICS

The surface horizon is generally 13-20 cm. thick. The colours vary in chroma from 3to 5, value from 2 to 4 in hues of 10YR. The effervescence is weak to strong.

LAND USE AND MANAGEMENT

These soils are generally under cultivation of paddy. The fields are properly managed and the area is covered under irrigation. Occasionally, the soils are also found to occur in waste lands of foot-hill slopes where scanty vegetation is encountered here and there.

ASSOCIATED SOILS

These soils are associated with Loamy Skeletal Udic Haplustalfs in H31 and Fine Vertic Ustochrepts in B12.

Pedon No. 8

FINE MIXED HYPERTHERMIC VERTIC HAPLAQUEPTS

These soils are very deep, very dark greyish brown to dark greyish brown, calcareous, moderately well to poorly drained and developed over the valleys of buried pediment plain. The soils are derived from alluvium and show cracks that are 1-2 cm. wide and go deep unto 50 cm. from the surface of the profile.

Location	:	About 2 km. west of Bamdevpalli Banswada/ Bodhan road. 100 metres inside
Physiographic position	:	Valley of buried pediment plain.
Drainage	:	Moderately well to poorly drained.
Land use	:	Cultivation
Parent material	:	Alluvium
Sampled by	:	S.C. Dhar, V.K.B. Sawant, S. Gopi
Profile No.	:	59
Date of examination	:	23.2.1985

Horizon	Depth (cm.)	Description
Ap	0-10	Brown (10YR 5/3) moist; dark greyish brown (10 YR 4/2) dry; silty clay loam, with few, fine distinct brown (7.5 YR 5/4) mottles; weak medium subangular blocky structure; very hard (dry); friable (moist); slightly plastic (wet); many fine and common medium pores; common fine and medium roots; moderate effervescence; clear smooth boundary.
Ap2	10-28	Dark greyish brown (10YR 3/2) moist; sandy clay with few fine distinct brown (7.5YR 5/4) mottles; weak medium subangular blocky structures; very hard (dry); firm (moist); very sticky and very plastic (wet); many very fine pores; very few very

		fine and medium roots; moderate effervescence; clear smooth boundary.
B1	28-65	Very dark greyish brown (10YR 3/2) moist; sandy clay; with few fine distinct grey (2.5Y 5/0) mottles; massive structure; firm (moist); very sticky and very plastic (wet); few very fine pores; very few very fine roots; weak effervescence; gradual smooth boundary.
B22	65-110	Very dark greyish brown (10YR 3/2) moist; sandy clay with few fine distinct grey (2.5Y 5/0) mottles; massive structure; firm (moist); very sticky and very plastic (wet); few very fine pores; moderate effervescence; gradual smooth boundary.
B22	110-130	Very dark greyish brown (10 YR 3/2) moist; sandy clay; massive structure; firm (moist), very sticky and very plastic (wet); few very fine pores; moderate effervescence.

LAND USE AND MANAGEMENT

These soils are cultivated for paddy and sugarcane. The fields are levelled and bunded.

ASSOCIATED SOILS

These soils are associated with Fine Vertic Ustochrepts in B2.

Pedon No. 9

FINE MONTMORILLONITIC HYPERTHERMIC TYPIC CHROMUSTERTS

These soils are very deep, dark brown to very dark greyish brown, calcareous, moderately well to poorly drained, derived from alluvium and found to occur on alluvial plain and buried pediment plain. The soils are cracking in nature and show the presence of slicknsides which are intersecting the ped faces. The cracks are 1-2 cm. wide and go deep more than 50 cm. from the soil surface.

Location	:	About 1/2 km. NE of Srinivas Camp.
Physiographic position	:	Buried pediment plain.
Land Use	:	Moderately well drained.
Parent material	:	Alluvium
Sampled by	:	S.C. Dhar, V.K.B. Sawant, S. Gopi
Profile No.	:	39
Date of examination	:	18.2.1985

Horizon	Depth (cm.)	Description
Ap	0-18	Very dark greyish brown (10YR 3/2) moist; greyish brown (10YR 5/2) dry; silty clay; weak medium subangular blocky structure; hard (Dry) ; friable (moist); very sticky and very plastic (wet); few fine and medium pores; few fine and many medium roots; moderate effervescence; clear smooth boundary.
AC	18-32	Very dark greyish brown (10YR 3/2) moist; silty clay, moderate medium subangular blocky structure; firm (moist); very sticky and very plastic (wet); common fine pores; few very fine roots; weak effervescence; gradual smooth boundary.

C1	32-62	Very dark (10YR 3/1.5) moist; silty clay; medium moderate angular blocky structure; firm (moist); very sticky and very plastic (wet); common fine pores; very few fine roots; weak effervescence; gradual smooth boundary.
C2	62-95	Dark greyish brown (2.5Y 4/2) moist; silty clay; medium moderate angular blocky structure; firm (moist); very sticky and very plastic (wet); few medium pores; weak effervescence; gradual smooth boundary.
C3	95-150	Dark greyish brown (2.5Y 4/2) moist; silty clay; medium moderate angular blocky structure; firm (moist), very sticky and very plastic (wet); few fine pores; weak effervescence.

RANGE IN CHARACTERISTICS

The surface layer is generally 15-20 cm. thick. Effervescence is moderate to strong. The chroma ranges from 3 to 4 and value from 1 to 3 in hue of 10YR and 2.5Y.

LAND USE AND MANAGEMENT

The soils are mostly under sugarcane cultivation and occasionally found on paddy cultivated areas. The fields are bunded and levelled.

ASSOICATED SOILS

The soils are associated with Fine Vertic Ustochrepts in B11.

Pedon No. 10

FINE MONTMORILLONITIC HYPERTHERMIC TYPIC PELLUSTERTS

These soils are very deep, very dark grey to very dark greyish brown, calcareous, moderately to poorly well drained derived from limestone material. These soils are cracking in nature and show the pressure slickensides which are intersecting in ped faces. Cracks are 1-2 cm. wide and go even below 50 cm. from the surface.

Location	:	About 1 km. south of Donkal village.
Physiographic position	:	Nearly level limestone landscape.
Drainage	:	Moderately well drained.
Land use	:	Cultivation
Parent material	:	Limestone
Sampled by	:	R.S. Yadava and D.R. Singh
Profile No.	:	73
Date of examination	:	25.2.1985

Horizon	Depth (cm.)	Description
Ap	0-10	Very dark grey (10YR 3.5/1) moist; grey (10YR 5/1) dry; clayey; fine medium subangular blocky; hard (dry), firm (moist) and very sticky and very plastic (wet); common very fine and fine pores; few fine and many very fine roots; weak lime reaction; clear smooth boundary.
AC	10-30	Very dark grey (10YR 3.5/1) moist; clayey; coarse subangular blocky; firm (moist), very sticky and very plastic (wet); few medium very fine and fine pores; few medium and many very fine roots; weak lime reaction, gradual smooth boundary.
C1	30-43	Very dark grey (10YR 3/1) moist; clayey; medium angular blocky structure with

		wedge shaped intersecting ped faces; firm (moist); very sticky and very plastic (wet); many fine pores few medium and common very fine roots; moderate lime reaction; gradual smooth boundary.
C2	43-70	Very dark grey (10YR 3/1) moist; clayey; medium angular blocky structure; wedge shaped intersecting ped faces; firm (moist); very sticky and very plastic (wet); many very fine pores; few very fine roots; weak lime reaction; gradual smooth boundary.
C3	70-125+	Very dark grey (10YR 3/1) moist; clayey; medium angular blocky structure; with intersecting wedge shaped peds; firm (moist); very sticky and very plastic (wet); many very fine pores; weak lime reaction.

RANGE IN CHARACTERISTICS

Surface layer is 10-20 cm. thick. Effervescence is weak. Chroma varies from 3 to 4 and value remains 1 in 10YR hue.

LAND USE AND MANAGEMENT

Soils are under paddy cultivation with palm trees. Bunding has been done to take good paddy crops.

ASSOCIATED SOILS

Fine Montmorillonitic Hyperthermic Typic Chromusterts.

Pedon No. 11

LOAMY SKELETAL MIXED HYPERTHERMIC UDIC HAPLUSTALFS

These are very deep soils characterized with dark reddish brown, well to excessively drained, no lime reaction and developed in granitic pediment plain. Soils are *in situ* in formation

Location	:	About 200 east of canal which intersects road after Ankspuram. (About 500 m. south of road crossing).
Physiographic position	:	Unterraced uncultivated pediment plain.
Drainage	:	Well drained.
Land use	:	Waste land
Parent material	:	Granitic gneiss
Sampled by	:	R.S. Yadava and D. R. Singh
Profile No.	:	10
Date of examination	:	12.2.1985

Horizon	Depth (cm.)	Description
Ap	0-17	Reddish brown (5YR 3.5/3); sandy loam; weak subangular blocky structure; very hard (dry); slightly sticky (moist); non-plastic (wet); few medium and many very fine and fine pores; many coarse roots; no lime reaction; clear smooth boundary.
B21t	17-28	Reddish brown (5YR 3.5/3.5); sandy loam/plus, weak subangular blocky structure; hard (dry); slightly sticky (moist); slightly plastic (wet); few medium and many very fine and fine pores; common thin clay cutans on ped faces and pores; few coarse and many medium roots; no lime reaction; abrupt wavy boundary.
B22t	28-88	Dark reddish brown (5YR 3/4), gravelly

		sandy clay loam; weak subangular blocky structure;very sticky (moist); plastic (wet); many very fine and fine pores; common patchy thin cutans on ped faces; few medium roots; no lime reaction; diffused smooth boundary.
B23t	88-115	Dark reddish brown (5YR 3/4); gravelly sandy clay loam plus; weak subangular blocky structure; sticky (moist); plastic (wet); few coarse & many very fine pores; common patchy thin cutans on ped faces and pores; few medium roots; no lime reaction; diffused irregular boundary.
C	115-155+	Dark reddish brown (5YR 3.5/4); gravelly sandy clay loam; weak subangular blocky structure; slightly sticky (moist) slightly plastic (wet); few medium and many fine pores; roots absent; no lime reaction.

RANGE IN CHARACTERISTICS

Reddish brown colour, deep soils, gravelly sandy loam to sandy loam texture, no lime reaction.

LAND USE AND MANAGEMENT

These soils are waste land and subject to overgrazing and moderate to severe erosion. Forestry measures should be adopted in this mapping unit.

ASSOCIATED SOILS

Fine Loamy Mixed Hyperthermic Udic Haplustalfs.

Pedon No. 12

LOAMY SKELETAL MIXED HYPERTHERMIC LITHIC USTORTHENTS

These soils are shallow to moderately deep, well drained with dark brown to reddish brown colour.

Physiography	:	Steep hill side slope
Drainage	:	Excessively drained.
Parent material	:	Granitic gneiss

Horizon	**Depth (cm.)**	**Description**
A1	0-18	Dark reddish brown colour (5 YR 3/2.5) moist; reddish brown (5YR 4/3) dry; gravelley sandy loam texture; weak fine subangular blocky; slightly hard (dry); very friable (moist); slightly sticky non-plastic (wet); few medium and coarse and many fine and very fine pores; common very fine and fine roots; no lime reaction; clear smooth boundary.
R	18	Lithic contact.

LAND USE AND MANAGEMENT

Sparse vegetation with grasses, which is subject to grazing.

INDEX

Aerial Photographs, 12, 13, 18
Agriculture, 9
Alfisols, 30
Alluvial Plain, 22, 34, 35
Argillic Horizon, 28
Automated Cartography, 92

Basement, 1
Base Saturation, 24
Buried Pediment, 4, 22, 31

Cambic Horizon, 31
Catenary Sequence, 23
Carbonate Removal, 22
Chromusterts, 35, 36, 37
Command Area, 1, 10
Conservation Measures, 53
Crop Rotation, 45
Cycle of Erosion, 4

Daccan Block, 1
Dot Grid Method, 19
Dykes, 28

Ecological Balance, 92
Ecoorder, 92
Environment, 20, 92
Epipedon, 22

Factors of Soil Formation
Final Cartography, 19
Final Legend, 18
Flood Plain, 22
Fluvial Process, 4

Gilgai Relief, 23, 36
Gneissic Complex, 25, 27
Granitic Intrusion, 1
Gross Cropped Area, 10

Haplaquepts, 31
Haplustalfs, 24, 30
Illuviation, 22
Infiltration-Runoff Ratio, 22
Inselbergs, 4, 28
In Situ, 28, 36

Land Evaluation, 44, 46, 47
Land Information System, 92
Land System, Facet, 12, 13
Land Qualities, 45, 46, 47
Land Suitability, 46, 47, 53, 60, 67, 71
Land Irrigability Classes, 79, 80
Land Use Planning, 44
Land Qualities, 46
Land Utilization Types (LUTs), 42, 45, 59, 60, 90
Leaching, 29
Levee and Back Swamp, 22
Legend, 15, 18
Levelling and Terracing, 30
Limestone Landscape, 22, 36, 37
Lithic Ustorthents, 24

Man Supporting Capacity, 24
Master Profile, 14, 17
Micro-Bioligical Activities, 22
Micro Relief Differences, 23

Mirror Stereoscope, 13
Montmorillomitic, 23, 36, 37, 40
Multispectral sensors, 92

Natural Resources, 7

Overlap-Forward and Lateral, 13

Paleustalfs, 24
Parcelling, 31, 34, 35, 36
Permeability, 80, 81
Pre-Field Interpretation, 14, 15
Penultimate Stage, 29
Phases of Soil, IX, 90
Photo Element, 18
Photo Index, 14
Photo-Interpretation, 12, 14
Preparation of Photographs, 14
Preservation of Environment, X

Quality & Quantity of Water,79

Rational Utilisation, X
Regulated markets, 11
Report Writing, 19
Repeatitive Coverage, 92
Resolution, 92
Residual Hills, 1
Rills and Gullies, 36

Sample Strips, 14, 16, 17
Secondary Plant Succession, 8
Semi-Detailed, IX, 12, 90
Slicken Sides, 23, 36, 37
Soil Irrigability Classes, 78
Soil Potentiality, IX
Soil Survey Interpretation, 12, 44
Soil Taxonomy, IX, 14
Stoniness and Rockiness, 23, 24
Suitability, 21, 47, 55, 60, 67, 71, 77
Summer Monsoon, 5, 35
Synoptic View, 21
Systematic Photo-Interpretation, 12, 21

Telengana Pediplain, 1
Terrain Morphology, 18
Thematic Mapping, 12, 21, 92

Uncontrolled Mosaic, 14
Ustochrepts, 23, 27, 28, 31, 34, 36
Ustorthents, 24

Water Holding Capacity, 80, 81
Watershed Management, 44

Zoom Transferoscope, 19